WITHDRAWN FROM
TSC LIBRARY

Renewable Energy in Nontechnical Language

Renewable Energy in Nontechnical Language

by Ann Chambers

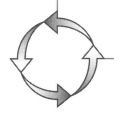

PennWell®

Copyright © 2004 by
PennWell Corporation
1421 South Sheridan/P.O. Box 1260
Tulsa, Oklahoma 74112/74101

800.752.9764
+1.918.831.9242
sales@pennwell.com
www.pennwellbooks.com
www.pennwell.com

Managing Editor: Kirk Bjornsgaard
Production Editor: Sue Rhodes Dodd
Book Designer: Wes Rowell
Cover Designer: Ken Wood

Library of Congress Cataloging-in-Publication Data

Chambers, Ann.
 Renewable Energy in Nontechnical Language / by Ann Chambers.-- 1st American ed.
  p. cm.
 ISBN 1-59370-005-9
 1. Renewable energy sources. I. Title.
 TJ808 .C47 2003
 333.79'4--dc22
 2003016041

All rights reserved. No part of this book may be reproduced, stored in a retrieval system, or transcribed in any form or by any means, electronic or mechanical, including photocopying and recording, without the prior written permission of the publisher.

Printed in the United States of America
 1 2 3 4 5 07 06 05 04 04

More PennWell Titles by Ann Chambers

Merchant Power: A Basic Guide

Natural Gas and Electric Power in Nontechnical Language

Power Primer: A Nontechnical Guide from Generation to End Use

Power Branding

Co-author of:

Distributed Generation: A Nontechnical Guide

Power Industry Desk Reference, available on CD or as e-book

Power Industry Dictionary

Power Industry Abbreviator

All titles are available at the PennWell Online Store
http://store.yahoo.com/pennwell.html
or call toll free *800-752-9764* to place an order or request a free catalog.

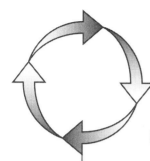

Contents

1. **Introduction** 1
   - Wind 4
   - Solar 4
   - Hydroelectric 5
   - Geothermal 5
   - Biomass 5
   - ***United States*** 7
     - Renewable futures 8
   - ***World Overview*** 11
     - Mexico 11
     - Western Europe 12
     - Industrialized Asia 13
     - Developing Asia 14
     - Central and South America 15
     - Eastern Europe/Former Soviet Union 16
     - Africa/Middle East 16

2. **Solar and PV** 19
   - PV 20
   - ***Solar Basics*** 21
     - Concentrating solar power 22
     - Passive solar heating and daylighting 23
     - Solar process heat 25

Market for PVs 26
Retail ready systems 30
Case Study: Engineering Firm Saves with Solar 32
System specifications 33
DOE Solar Buildings Program 34
Research and development 35
Case Study: Solar Patriot House Displays Green Design 36
Types of Solar Cells 40
Case Study: Santa Rita Jail Boasts Biggest Roof Installation ... 41
Case Study: Largest Solar Installation in the United Kingdom ... 42
Case Study: 4000 Solar Systems for Rural Telecom 44
Case Study: 3000 Solar Panels Save University $50,000 Annually ... 45
**Case Study: Peru Houses First Latin American
Solar Grid-Connect System** 47
Case Study: Fun in the Sun 48
Case Study: Learning From Light Brightens Schools 49
Case Study: Staying Green at Hybrid Parker Ranch 50
Equipment 52
Benefits 52
Future views 52
Case Study: HCPV—Catching Rays in Arizona 54
How it works 56
Efficiency measures 57
Case Study: Australian Solar Tower Breaking Records 58
Financing 59
Design and construction 60

3. Wind Energy 61
Types of turbines 62
U.S. wind resources 63

	Falling costs	64
	Industry statistics	66
	Merchant potential	71
	New wind technologies	71
	Siting hurdles	71
	Unique considerations	72
	Guidelines	73
	Land use	74
	For the birds	76
	Sound concerns	77
	Visionary planning	78
	Case Study: Big Spring Keeps on Turning	80
	Offshore aesthetics	82
	Capacity and output	86
	Transmission issues	88
	Case Study: Winds of Change in Italy	91
	Case Study: Arklow Bank Project Siting 200 Turbines Offshore	93
	Case Study: Brazil's Hydro Crisis Unleashes Wind Demand	95
4.	**Bioenergy**	97
	Case Study: Landfill Gas-to-Energy Powers Iowa	99
	Biomass potential	102
	Case Study: PGE Makes Green Power with Cows	106
	Biomass for heat and power	107
	Back for the future	108
	Biogas potential	109
	Biopower basics	111
	Transportation applications	114
	Ethanol	114

Biodiesel 115
   Biodiesel market 116
Case Study: Savannah River Switches to Biodiesel 117
Case Study: Florida, Alabama Utilities Laud Biodiesel 118
   Other biobased products 120
   Biomass at work 122
World Views 124
Case Study: Biomass Use in Bangladesh 126

5. Geothermal Energy 129
   Steam 131
   High-temperature water 131
   Moderate-temperature water 132
   Power production 133
   The future of geothermal power 135
   DOE developments 137
   Principal research and development thrusts 137
Case Study: New Nevada Campus to be Fully Powered by Geothermal Energy 139
Case Study: Calpine and the Geysers 142
   Geothermal outlook 144

6. Hydroelectric 147
   North America 151
   United States 151
   Canada 152
   Central and South America 155
Case Study: Energy Crisis in Brazil—Implications for Hydropower 155
Hydropower Basics 157
   Conventional 158
   Pumped storage 158
Types of Hydropower 160

    Impoundment 160
    Diversion 160
    Storage facilities 160
    Sizes of hydropower plants 161
    Turbine technologies 161
Environmental Issues and Mitigation 163
    Development forecast 164
    Hydro licensing 166
Legislative and Regulatory Considerations 167
    National Environmental Policy Act of 1969 167
    Clean Water Act of 1977 168
    Fish and Wildlife Coordination Act of 1934 168
    National Historic Preservation Act of 1966 168
    Wild and Scenic Rivers Act of 1968 168
    Endangered Species Act of 1973 169
    Coastal Zone Management Act of 1972 169
    Americans with Disabilities Act of 1990 169
Case Study: Hydropower in Hawaii 170
    Kauai 170
    Maui 171
    Hawaii 171

7. Fuel Cells 173
Three Main Fuel Cell Types 176
    PAFC 177
    MCFC 178
    SOFC 179
    Polymer exchange membrane (PEM) and alkaline 180
    Race to commercialization 182
    Gaining popularity 184
    Reliability 187
    The hydrogen economy 187

      Geography 188
      Forward view 190
    Case Study: Wabash River and Clean Coal Fuel Cell Trials 191
    Case Study: European Residential Promise 193
      H_2 issues 195

Appendix A: Industry Contact List 197
Domestic 197
International 202

Appendix B: Renewable Energy Glossary 205

Index 233

Introduction

Renewable energy sources are those that will replenish themselves—the tide, water flowing downhill in a river, the wind, and the sun. As the technologies to use these natural energy resources evolve, the expense of using renewable energy to generate electricity is coming more in line with the traditional fuels. Emissions constrictions on fossil fuels, research and development efforts, and government subsidies are bringing this niche market toward the mainstream in the United States. Government subsidies also help bring the costs more in line.

Renewables are also gaining ground in the marketing arena. They supply what is often referred to as *green power*, and the U.S. public has been embracing green-power programs offered by utilities as a way to please environmentally conscious customers (Figs. 1–1 and 1–2).

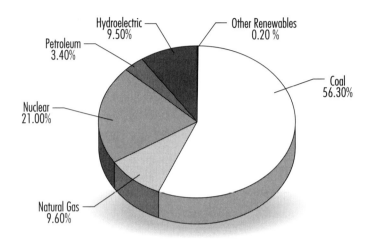

FIG. 1–1 U.S. ELECTRIC POWER GENERATION BY FUEL TYPE

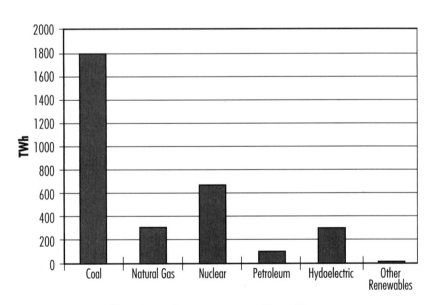

FIG. 1–2 GENERATION BY FUEL TYPE

¶ Renewable resources—solar, wind, geothermal, hydroelectric, biomass, and municipal solid waste—provide about 12% of the U.S. electricity supply. Hydroelectric resources alone provide almost 10% of this, although hydro is not really a contender as a distributed generation source. Biomass and municipal solid waste (MSW) together contribute a little more than 1%. All other renewable resources—including geothermal, wind, and solar—together provide less than 1% of the total (Figs. 1–3, 1–4, and 1–5).

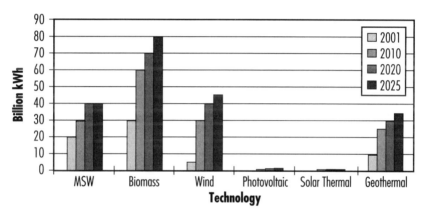

Fig. 1–3 Renewables Growth Projections

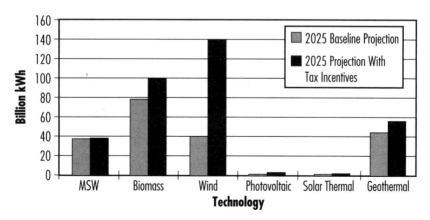

Fig. 1–4 Projections by Technology

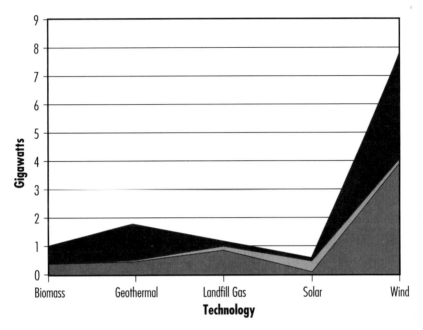

FIG. 1–5 PROJECTED RENEWABLE ADDITIONS THROUGH 2025

Wind

The wind contains lots of energy. Wind turbines use the wind to create electricity. Wind power creates no pollution and has very little impact on the land. Wind energy can be produced anywhere the wind blows consistently.

Solar

The sun's radiation is used directly to produce electricity in two ways. Photovoltaic (PV) systems turn sunlight into electricity directly. Solar thermal systems use the sun's heat to heat water, creating steam to turn a turbine and generator.

Hydroelectric

Dams provide electricity by guiding water down a chute and over a turbine at high speed. Small hydropower facilities are considered renewable energy resources. Large dams are not. Hydropower does not produce any air emissions, but large dams have environmental issues such as flood control, water quality, and fish and wildlife habitat to deal with.

Geothermal

Geothermal energy is generated by converting hot water or steam from deep beneath the earth's surface into electricity. Geothermal plants cause very little air pollution and have minimal impacts on the environment.

Biomass

Organic matter, called *biomass*, can be burned in an incinerator to produce energy. In some facilities, the biomass is converted into a combustible gas, allowing for greater efficiency and cleaner performance. Biomass comes from a variety of sources, including agricultural, forestry, or food-processing byproducts, as well as gas emitted from landfills.

Many renewable resources are relative newcomers to the electric power market. In particular, electricity generation using geothermal, wind, solar, and MSW resources has had its greatest expansion since the 1980s. This is the result of significant technological improvement, the implementation of favorable governmental policies, and the reaction to the increasing costs of using fossil and nuclear fuels. Wind is the fastest growing renewable and resource and likely the most viable as a distributed generation resource. Use of renewable resources for electricity generation has also been encouraged as less environmentally damaging than fossil fuels. Because renewable energy is available domestically, renewable resources are viewed by some as more secure than imported fossil fuels.

The use of renewable resources other than hydroelectricity is increasing, particularly wind power, which has enjoyed rapidly developing technology for efficiency improvements and government incentives bringing down the expense. Conventional hydroelectric power, the mainstay of renewable resources in electric power today, is unlikely to enjoy rapid growth under current expectations, even if more favorable regulatory policies emerge. The lack of many additional large sites for hydroelectric facilities limits the potential for hydroelectric power growth (*see* Tables 1–1 and 1–2).

If renewable resources are to provide a greater share of the U.S. electricity supply, costs of using them will need to decline relative to alternatives. In some cases, such as wind and solar thermal generation, small improvements in generating costs may significantly increase their market penetration. In other cases, such as PV and most forms of geothermal power, large cost reductions are needed to spur greater market penetration.

TABLE 1–1
EXISTING CAPACITY AND PLANNED ADDITIONS AT U.S. ELECTRIC UTILITIES

Fuel	2000 No. of Units	2000 Capacity (MW)	2005 No. of Units	2005 Capacity (MW)
Coal	1,024	278,005	NA	NA
Petroleum	3,007	45,646	95	984
Natural Gas	2,068	129,516	280	47,549
Hydro–Pump Storage	135	17,480	NA	NA
Conventional Hydro	2,836	71,185	24	392
Nuclear	91	92,111	NA	NA
Waste Heat	61	4,584	13	1,817
Other Renewables*	128	889	NA	NA

*Includes geothermal, biomass, solar and wind
Source: DOE

TABLE 1–2
Existing Capacity and Planned Additions at U.S. Nonutilities

Fuel	2000 No. of Units	2000 Capacity (MW)	2005 No. of Units	2005 Capacity (MW)
Coal	599	59,196	7	2,928
Petroleum	600	10,969	22	3,236
Natural Gas	1,668	72,453	1,517	230,726
Other Gas	104	2,455	NA	NA
Petroleum/Natural Gas (Combined)	765	47,580	279	39,381
Hydroelectric	1,418	7,214	10	71
Nuclear	13	12,622	NA	NA
Solar	9	410	NA	NA
Wind	88	2,323	65	1,277
Wood	324	6,246	6	119
Waste	680	3,890	NA	NA
Other	25	545	NA	NA

Source: DOE

United States

Potential sites for hydroelectric dams have already been largely established in the United States, and environmental concerns are expected to prevent the development of any new sites in the future. U.S. conventional hydroelectric generation is expected to decline from 316 billion kilowatt-hours (kWh) to 304 billion kWh in 2020 as increasing environmental and other competing needs reduce the productivity of generation from existing hydroelectric capacity.

Other renewables are expected to account for 3.9% of all projected additions to U.S. generating capacity through 2020. Generation from geothermal, biomass, landfill gas, solar, and wind energy is projected to increase from 77 billion kWh in 1999 to 160 billion kWh in 2020.

Biomass (including cogeneration and co-firing in coal-fired power plants) is expected to grow from 38 billion kWh in 2000 to 64 billion kWh in 2020. Most of the increase is attributed to cogenerators, with a smaller amount from co-firing. Over the next 10 to 15 years, only a few new dedicated biomass plants are expected to be constructed.

High-output geothermal capacity could increase by 87% over the next two decades, to 5300 megawatt (MW), and provide almost 35 billion kWh of electricity generation by 2020. This will depend, however, on the success of several new untested sites.

Wind capacity in the United States is projected to grow by nearly 300% by 2020 to about 9100 MW.

Renewable futures

Energy is regulated from the state and the national levels, but one of the most promising regulatory trends for renewable energy is the state-based move toward mandated renewable portfolio standards (RPS). Federal efforts at a nationwide standard have fallen short, but states are individually moving forward.

Texas passed its RPS bill in 1999 when it restructured the power industry to introduce retail electric choice. Many Texas consumers are now paying extra—voluntarily—to buy power from one of many wind farms built in Texas over the past few years. If the trend continues, Texas could well pass California as the nation's top wind energy state. Texas has a goal of an additional 2000 MW of renewable generation by 2009.

Arizona has instituted green-pricing programs, net metering, and a limited renewable portfolio standard.

California, which captured headlines and national attention with rolling blackouts, price spikes, and scandals a couple of years ago, has the most green-power plants. The state offers a solar and wind energy tax credit, covering up to 15% of a qualifying system, solar property tax exemptions, green-pricing programs through area utilities, net meter, and state loan programs for renewable energy systems.

Nevada likewise offers property tax exemptions, green pricing systems, net metering, and an enacted RPS. Utilities must meet a renewable energy requirement that started at 5% in 2003, escalating to 15% by 2013 (*see* Table 1–3).

At this writing, 13 states had RPS and another 3 had stated renewables goals. States with enacted RPS legislation include: Arizona, California, Connecticut, Hawaii, Illinois, Iowa, Maine Massachusetts, Minnesota, Nevada, New Jersey, New Mexico, Texas, and Wisconsin. To see a map of the latest RPS states from the Database of State Incentive for Renewable Energy (DSIRE), log on to www.dsireusa.org.

TABLE 1–3
RENEWABLE ELECTRIC POWER NET GENERATION

State	Geothermal	Hydro	MSW/Landfill Gas	Biomass	Other Solar	Wind	Wood/WoodWaste
Alabama		5,817					142
Alaska		1,001					
Arizona		8,354					
Arkansas		2,370					
California	12,308	38,325	1,808	391	493	3,518	2,454
Colorado		1,454		7			
Connecticut		526	1,957	197			
Delaware			19				
District of Columbia							
Florida		87	3,030	404			401
Georgia		2,459	7				
Hawaii	262	43	350				17
Idaho		10,997					39
Illinois		142	611	266			
Indiana		588	88				
Iowa		904	71				
Kansas		15					
Kentucky		2,325					
Louisiana		532		64			
Maine		2,295	242	101			
Maryland		1,733	628				
Massachusetts		1,053	2,059	1			
Michigan		1,401	690	64			1,045
Minnesota		684	773				725

Table 1–3 (Continued)

State	Geothermal	Hydro	MSW/Landfill Gas	Biomass	Other Solar	Wind	Wood/WoodWaste
Mississippi							
Missouri		600	73				
Montana		9,623					
Nebraska		1,501		7			
Nevada	1,371	2,429					
New Hampshire		1,244	244				
New Jersey		14	1,350				
New Mexico		221		8			
New York		24,819	1,992	1		10	383
North Carolina		2,192	98				372
North Dakota		2,123					
Ohio		583	27				44
Oklahoma		2,277					
Oregon		38,116	95			67	268
Pennsylvania		2,290	1,734	4		10	194
Rhode Island		5	115				
South Carolina		1,533					
South Dakota		5,716					
Tennessee		5,876	29				1
Texas		828	60		1	492	
Utah	152	746	9				
Vermont		1,201				12	330
Virginia		699	106	2			335
Washington		80,160	205	29			429
West Virginia		698		14			
Wisconsin		1,754	210	76		3	163
Wyoming		1,011					246

Source: EIA

World Overview

Mexico

In Mexico there are limited plans to expand the renewable energy resource base (Fig. 1–6). Mexico has made some moves toward increasing the development of geothermal resources, including studies by the state-owned Comisión Federal de Electricidad (CFE) and a government pledge to invest some $31 million in geothermal energy.

There has been little activity in wind power development in Mexico, although by some estimates, Mexico has wind resources that could support the installation of up to 5000 MW of wind power capacity. The country has about 3 MW of installed wind capacity but has not added any new capacity since 1998.

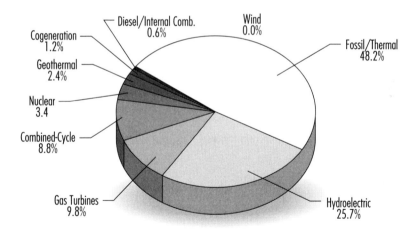

FIG. 1–6 MEXICAN GENERATION MIX

Western Europe

Expansion of renewable energy sources in Western Europe is expected to be mostly in the form of nonhydroelectric renewables. Most potential hydroelectric resources have already been developed in the region, and there are few plans to extend hydropower capacity over the next two decades.

Among the other forms of renewable energy, wind has made the greatest gains over recent years and will probably contribute to much of the future growth in renewable energy use.

The European Union (EU) has moved to increase the penetration of renewables in the European energy mix. In 2001, the European Parliament approved a Renewables Directive that would require the EU to double the renewable share of total energy consumption by 2010. According to the new law, the share of total inland energy consumption met by renewable energy resources will have to increase to 12% in 2010, from an estimated current level of about 6%. Furthermore, the share of electricity demand met by renewables will have to increase to 22%, from about 14% now.

Individual European countries have been implementing various strategies to increase use of renewables. The United Kingdom introduced a *renewables obligation*, which required electricity suppliers to derive 3% of their electricity from renewable resources in 2002, with requirements escalating to 10% in 2011.

Germany's Gesetz für den Vorrang Erneuerbarer Energien law was enacted on April 1, 2000. It requires that electricity grid operators give *priority access to all renewable energy* and sets fixed rates for each renewable (the cost is passed to the consumer).

France has also set rates for renewable energy in the wholesale market to ensure that a planned installation of 10,000 MW of wind power occurs by 2010.

In contrast to the German and French strategies of rate setting, the government of the Netherlands uses *green certificates* to create a market for renewables. Generators are given green certificates for their renewable power production that provide tax credits and can be traded. The resulting tax savings or the earnings made from the sale of the certificates are supposed to allow renewable generators to sell more of their power in the market.

In the past, Denmark has required utilities to allow private renewable energy producers access to the grid and has required utilities to pay the producers a percentage of their production and distribution costs. Now the Danish government is also introducing renewable energy certificates, similar to the Dutch scheme.

Of all the renewable energy sources, wind is the most promising in Europe. Germany, Spain, and Denmark have been among the world's top wind capacity installers in recent years, and recently both Italy and the United Kingdom have seen sharp increases in wind power capacity installations. Even some European countries that have been slow in developing wind programs heretofore are beginning to make plans for expanding this renewable energy source. Offshore wind is allowing European countries not having the land area to devote to wind turbines a chance to begin exploiting wind energy.

Industrialized Asia

The countries of industrialized Asia (Australia, Japan, and New Zealand) have markedly different electricity energy mixes. Japan is the only one of the three countries with a nuclear generation program, supplying one-third of its electricity from nuclear power plants. Hydroelectricity and other renewable energy sources supply about 12% of Japan's electricity. Renewables also account for about 10% of Australia's electricity supply, and thermal generation (predominantly coal) accounts for nearly 90%. In contrast, renewable energy sources provide 73% of New Zealand's electricity supply.

Between 1999 and 2020, the use of hydroelectricity and other renewables is projected to increase by 1.4% per year in the region of Australasia (including Australia and New Zealand, along with the U.S. territories). Much of this modest increase is expected to be in the form of nonhydroelectric renewables, most notably wind.

Developing Asia

Support for the construction of large-scale hydroelectric dams remains strong in many countries of developing Asia. Large-scale hydropower projects in China, India, Malaysia, and Vietnam, among others in the region, are expected to provide most of the predicted 4.3% annual growth in renewable energy consumption worldwide.

There are more modest efforts to increase nonhydroelectric renewable energy use, primarily wind and solar, in China, India, and other developing Asian countries. The projects are often aimed at reaching small, rural communities that would otherwise not have access to electricity through the national grid.

In China, work progresses more or less on schedule on the 18,200 MW Three Gorges Dam project, the largest hydroelectric project in the world. The dam is being built on China's Yangtze River. It is scheduled to be fully operational by 2009. The Three Gorges Dam project has encountered problems with accusations of corruption, and there have been difficulties in relocating the estimated 1.13 million residents who will have to move before the dam's reservoir can be flooded. Since 1993, more than 350,000 residents have been relocated. Beyond the expansion of large-scale hydropower, several other projects are underway to develop China's other renewable resources, notably, wind and solar.

Vietnam also proposes expanding its large-scale hydroelectric power over the next several years. In 2001, Vietnam's National Assembly approved construction of the 3,600 MW Son La hydropower project to be constructed on the Da River, about 200 miles west of Hanoi. The project is the subject of some dispute, even among members of the National Assembly, because it has been sited for an area known to have frequent seismic disturbances, and is opposed by human rights activists because it would require the relocation of up to 700,000 people, mostly of ethnic minorities. Estimates for the cost of constructing Son La, which is scheduled for completion in 2016, have run as high as U.S. $5.1 billion. Proponents of the project have argued that it is needed to help improve Vietnam's electricity fuel mix, reduce flood damage, and improve irrigation in the Red River Delta.

Central and South America

Hydroelectricity is an important source of electricity generation in Central and South America. In Brazil, the region's largest economy, hydropower typically supplies more than 90% of the country's electricity generation. As a result, drought can have a devastating impact on electricity supply, and many countries of Central and South America are initiating projects to diversify the mix of electricity supply. Much of the diversification will consist of adding natural gas-fired electricity capacity to reduce dependence on hydropower.

As a result, although there is some projected growth in the use of hydroelectric and other renewable resources in the forecast, it is expected to be much less than the growth in natural gas consumption. Demand for hydroelectricity and other renewables in Central and South America is expected to jump by only 1.4% annually through 2020.

Many Central and South American countries are attempting to address the problem of getting electricity to remote, rural areas. Costa Rica has one of the most ambitious programs for renewable energy in Latin America. The country instituted a policy mandating that by 2025 all forms of energy consumed in the country be derived from renewable sources.

In Argentina, the government and the World Bank are implementing a project that is to provide electricity to roughly 70,000 rural households and 1100 provincial public service institutions, principally through the use of renewable energy systems. Total cost of the project has been estimated at $120.5 million. Energy sources will be principally PV and wind, with biomass used to make up any shortfalls. Argentina has expressed a particular interest in developing its wind resources. The country has passed legislation that requires all utilities to purchase wind power if it is available. This should help cover the costs of building the necessary transmission infrastructure from the wind turbines to the power distributors. Further, with approval of the Argentine government, Spanish companies Endesa and Elecnor are developing 3000 MW of wind energy capacity, to be completed by 2010.

Eastern Europe/Former Soviet Union

There are only a few plans to expand the use of renewable resources in the countries of Eastern Europe and the Former Soviet Union (FSU). In general, renewables are not competitive in the FSU, where fossil fuel resources are abundant and demand for clean forms of electricity can be met with cheaper natural gas–fired capacity. FSU renewable energy demand is projected to increase by 0.8% per year. In Eastern Europe, the growth rates projected for hydroelectricity and other renewables are four times those for the FSU at 3.4% per year, reflecting the relatively small amount of renewable capacity currently installed in the region.

Africa/Middle East

In Africa and the Middle East, hydroelectricity and other renewable energy sources have not been widely established, except in a few countries. In the Middle East, only Turkey and Iran have extensively developed their hydroelectric resources. In Africa, Egypt and Congo (Kinshasa) have the largest volumes of hydroelectric capacity. Other countries, including Ivory Coast, Kenya, and Zimbabwe, are almost entirely dependent on hydropower for their electricity, not because they have extensively developed hydropower resources but rather because of a lack of development of electricity infrastructure.

There have been several recent advances in the development of nonhydroelectric renewable energy projects in Africa. Morocco continues its pursuit of wind power. The state-owned utility Office Nationale d'Electricité (ONE) is planning 200 MW of wind power at Tangiers and Tarfaya. The country's first wind power plant, the 50-MW Koudia al-Baida, began operating in May 2000 and is generating an estimated 200 million kWh of electricity annually.

South Africa's energy market is almost entirely supplied by state-owned Eskom, which is the fifth largest electric utility in the world. Of Eskom's almost 40,000 MW of capacity, more than 35,000 MW is coal fired. However, adding green power and trimming the coal use is very important for the country, which has serious pollution problems. Pollution from fossil fuels is one of the biggest killers of children under five in the country, and the government is beginning to step in with motivation for fuel diversification. Eskom does have one nuclear plant at Koeberg supplying 1840 MW, two gas turbine plants totaling 342 MW, and six hydroelectric plants adding 66 MW to its mix. Eskom also has two pumped storage stations capable of generating 1400 MW. South Africa is working toward a goal of 10,000 MW of green energy by 2012. This would total about 2% of the country's demand. The government is also working to privatize at least part of the power sector.

Egypt also has made some advances in wind power and the Egyptian New and Renewable Energy Authority (affiliated with the state-owned Egyptian Electricity Holding Company) hopes that wind will supply some 600 MW of electricity capacity to the national grid by 2007.

In the Middle East, much of the new development in renewable energy, particularly hydroelectricity, is centered in Turkey. The country has ambitious plans to construct a system of 21 dams and 19 hydroelectric plants, called the Southeast Anatolia Project. It is a joint hydroelectric power and irrigation project. Upon completion, the $32 billion project will have a combined installed capacity of 7500 MW. Several of the hydropower facilities are complete and more are under construction.

2

Solar Power and PVs

Each day more solar energy falls to the earth than the total amount of energy the planet's approximately 6 billion inhabitants would consume in 26 years. While it's neither possible nor necessary to use more than a small portion of this energy, we've hardly begun to tap the potential of solar energy.

Although every location on earth receives sunlight, the amount received varies greatly depending on geographical location, time of day, season, and light. The southwestern United States is one of the world's best areas for sunlight. This desert region receives almost twice the sunlight of other regions of the country. Solar energy systems use either solar cells or some form of solar collector to generate electricity or heat for homes and buildings. The primary solar energy technologies for power generation are PVs and thermal systems.

PV devices use semiconductor material to directly convert sunlight into electricity. Solar cells have no moving parts. Power is produced when sunlight strikes the semiconductor material and creates an electric current. Solar cells are used to power remote homes, satellites, highway signs, water pumps, communication stations, navigation buoys, streetlights, and calculators (Fig. 2–1).

FIG. 2–1 A PV INSTALLATION NEAR HOT SPRINGS
(COURTESY OF SHELL SOLAR)

PV

Solar thermal systems generate electricity with heat. Concentrating solar collectors use mirrors and lenses to concentrate and focus sunlight onto a receiver mounted at the system's focal point. The receiver absorbs and converts sunlight into heat. The heat is then transported to a steam generator or engine where it is converted into electricity.

Solar energy technologies offer a clean, renewable, and domestic energy source. The generating systems powered are also modular so they can be constructed to meet any size requirement and are easily enlarged to meet changing energy needs.

Solar energy technologies have made huge technological and cost improvements, but except for certain niche markets like remote power applications, they are still more expensive than traditional energy sources. Researchers continue to develop technologies that will make solar energy technologies more cost competitive.

Solar Basics

Solar cells convert sunlight directly into electricity using semi-conducting materials similar to those used in computer chips. When sunlight is absorbed by these materials, the solar energy knocks electrons loose from their atoms, allowing the electrons to flow through the material to produce electricity.

This process of converting light (photons) to electricity (voltage) is called the PV effect.

Solar cells are typically combined into modules that hold about 40 cells, then about 10 of the modules are mounted in PV arrays that can measure up to several meters on a side.

These flat-plate PV arrays can be mounted at a fixed angle facing south, or they can be mounted on a tracking device that follows the sun, allowing them to capture the most sunlight over the course of a day. About 10 to 20 PV arrays can provide enough power for a household. For large electric utility or industrial applications, hundreds of arrays can be interconnected to form a single, large PV system.

Thin film solar cells use layers of semiconductor materials only a few micrometers thick. Thin film technology has made it possible for solar cells to now double as rooftop shingles, roof tiles, building facades, or the glazing for skylights or atria. The solar cell version of such items as shingles offers the same protection and durability as ordinary asphalt shingles.

Some solar cells are designed to operate with concentrated sunlight. These cells are built into concentrating collectors that use a lens to focus the sunlight onto the cells. This approach has both advantages and disadvantages compared with flat-plate PV arrays. The main idea is to use very little of the expensive semiconducting PV material while collecting as much sunlight as possible. Because the lenses must be pointed at the sun, the use of concentrating collectors is limited to the sunniest parts of the country. Some concentrating collectors are designed to be mounted on simple tracking devices, but most require sophisticated tracking devices, which further limit their use to electric utilities, industries, and large buildings.

The performance of a solar cell is measured in terms of its efficiency at turning sunlight into electricity. Only sunlight of certain energies will work efficiently to create electricity, and much of it is reflected or absorbed by the material making up the cell.

Thus, a typical commercial solar cell has an efficiency of 15%—about one-sixth of the sunlight striking the cell generates electricity. Low efficiencies mean that larger arrays are needed, and that means higher cost. Improving solar cell efficiencies while holding down the cost per cell is an important goal of the PV industry, National Renewable Energy Laboratory (NREL) researchers, and other U.S. Department of Energy (DOE) laboratories, and they have made significant progress. The first solar cells, built in the 1950s, had efficiencies of less than 4%.

Concentrating solar power

Many power plants today use fossil fuels as a heat source to boil water. The steam from the boiling water rotates a turbine, which activates a generator producing electricity. However, a new generation of power plants, with concentrating solar power systems, uses the sun as a heat source. There are three main types of concentrating solar power systems—parabolic-trough, dish/engine, and power tower.

Parabolic-trough. Parabolic-trough systems concentrate the sun's energy through long rectangular, curved (U-shaped) mirrors. The mirrors are tilted toward the sun, focusing sunlight on a pipe that runs down the center of the trough. This heats the oil flowing through the pipe. The hot oil then is used to boil water in a conventional steam generator to produce electricity.

Dish/engine. A dish/engine system uses a mirrored dish (similar to a very large satellite dish). The dish-shaped surface collects and concentrates the sun's heat onto a receiver, which absorbs the heat and transfers it to fluid within the engine. The heat causes the fluid to expand against a piston or turbine to produce mechanical power. The mechanical power is then used to run a generator or alternator to produce electricity.

Power tower. A power tower system uses a large field of mirrors to concentrate sunlight onto the top of a tower, where a receiver sits. This heats molten salt flowing through the receiver. The salt's heat is used to generate electricity through a conventional steam generator. Molten salt retains heat efficiently, so it can be stored for days before being converted into electricity. That means electricity can be produced on cloudy days or even several hours after sunset. There is a case study later in this chapter detailing the largest, most innovate power tower currently under construction in Australia.

Passive solar heating and daylighting

Step outside on a hot, sunny summer day, and you'll feel the power of solar heat and light. Today, many buildings are designed to take advantage of this natural resource through the use of passive solar heating and daylighting.

The south side of a building always receives the most sunlight. Therefore, buildings designed for passive solar heating usually have large, south-facing windows. Materials that absorb and store the sun's heat can be built into the sunlit floors and walls. The floors and walls will heat up during the day and slowly release heat at night when the heat is needed most. This passive solar design feature is called direct gain.

Other passive solar heating design features include sunspaces and trombe walls.

Sunspace. A sunspace (which is much like a greenhouse) is built on the south side of a building. As sunlight passes through glass or other glazing, it warms the sunspace. Proper ventilation allows the heat to circulate into the building.

Trombe wall. A trombe wall is a very thick, south-facing wall, which is painted black and made of a material that absorbs a lot of heat. A pane of glass or plastic glazing, installed a few inches in front of the wall, helps hold in the heat. The wall heats up slowly during the day. As it cools gradually during the night, it gives off its heat inside the building.

Many of the passive solar heating design features also provide daylighting. Daylighting is simply the use of natural sunlight to brighten up a building's interior. To lighten up north-facing rooms and upper levels, a clerestory—a row of windows near the peak of the roof—is often used along with an open floor plan inside that allows the light to bounce throughout the building.

Of course, too much solar heating and daylighting can be a problem during the hot summer months. Fortunately, there are many design features that help keep passive solar buildings cool in the summer. For instance, overhangs can be designed to shade windows when the sun is high in the summer. Sunspaces can be closed off from the rest of the building. Also, a building can be designed to use fresh-air ventilation in the summer.

Solar process heat

Commercial and industrial buildings may use the same solar technologies—PVs, passive heating, daylighting, and water heating—that are used for residential buildings. These nonresidential buildings can also use solar energy technologies that would be impractical for a home. These technologies include ventilation air preheating, solar process heating, and solar cooling.

Many large buildings need ventilated air to maintain indoor air quality. In cold climates, heating this air can use large amounts of energy. A solar ventilation system can preheat the air, saving energy and money. This type of system typically uses a transpired collector, which consists of a thin, black metal panel mounted on a south-facing wall to absorb the sun's heat. Air passes through the many small holes in the panel. A space behind the perforated wall allows the air streams from the holes to mix together. The heated air is sucked out from the top of the space into the ventilation system.

Solar process heating systems are designed to provide large quantities of hot water or space heating for nonresidential buildings. A typical system includes solar collectors working along with a pump, a heat exchanger, and/or one or more large storage tanks.

The two main types of solar collectors used—an evacuated-tube collector and a parabolic-trough collector—can operate at high temperatures with high efficiency. An evacuated-tube collector is a shallow box full of many glass, double-walled tubes and reflectors to heat the fluid inside the tubes. A vacuum between the two walls insulates the inner tube, holding in the heat.

Parabolic troughs are long, rectangular, curved (U-shaped) mirrors tilted to focus sunlight on a tube, which runs down the center of the trough. This heats the fluid within the tube.

The heat from a solar collector can also be used to cool a building. It may seem impossible to use heat to cool a building, but it makes more sense if you just think of solar heat as an energy source. Your familiar home air conditioner uses an energy source—electricity—

to create cool air. Solar absorption coolers use a similar approach, combined with some very complex chemistry, to create cool air from solar energy. Solar energy can also be used with evaporative coolers to extend their usefulness to more humid climates, using another bit of chemistry called desiccant cooling, which is dehumidification through chemical means to ease the load on air conditioning equipment.

Market for PVs

According to some estimates, there is more than 1 gigawatt (GW) of PV power operating worldwide, with more than 350 MW of that added during 2002. Much of that power is being used as a distributed generation resource without connecting to the grid. PV is often sited near its point of use, often in remote locations that are too difficult or too expensive to serve through grid power.

However, future use of PV will likely involve more applications that are grid connected to supplement or back up grid power. In recent years, grid connected applications have actually been growing a bit faster than off-grid installations.

City Public Service of San Antonio's 75,000-square foot Northside Customer Service Center has been transformed into a model for renewable energy usage and energy-efficient design. The center features one of the largest combined-capacity, solar power roof panel systems in the United States, a natural gas chiller-heater and humidity-control system, a rainwater reclamation system, and recycled building materials (Fig. 2–2).

FIG. 2–2 CITY PUBLIC SERVICE OF SAN ANTONIO'S NORTHSIDE CUSTOMER CENTER (PHOTO COURTESY OF CITY PUBLIC SERVICE OF SAN ANTONIO)

The off-grid market has been growing at 15% to 20% annually and the grid-connected PV markets have jumped to a 30% annual growth rate.

Electric Power Research Institute (EPRI) studied PV systems in 2002, finding that many grid-connected systems are being built despite having a higher cost than conventional power.

Eight innovative PV leaders were studied including the following.

- **Sacramento Municipal Utility District (SMUD).** For more than a decade, SMUD has supported PV in its service territory, including its PV Pioneer I, a program allowing residential customers to pay a small additional monthly cost to support a SMUD-owned PV system.
- **Shea Homes.** This home developer is building 100 single-family homes, each with a 1.2 kW PV system as a standard feature.
- **Niagara Mohawk.** The utility installed a 101 kW PV system with battery storage at a utility feeder to provide whole-building uninterruptible power supply.
- **Gardner, Massachusetts.** Installation of 30 2.2 kW residential PV systems and five commercial systems in Gardner is concentrated on a 13.8 kV feeder.
- **Arizona Public Service (APS).** EPRI studied APS's ground-mount systems.
- **Mauna Lani Hotel.** The hotel has horizontally mounted PV panels on the building roof, providing 80 kW of power while reducing building heat gain.
- **Stelle, Illinois.** One-third of the homes in this small community have installed PV systems.
- **Guerilla Solar.** A small residential PV system that was connected to the grid without any permitting or utility involvement because the homeowner decided that working with the utility would be too troublesome.

The main observations revealed in the EPRI study included the following:

Grid-connected PV. The most notable observation was the surprising amount of grid-connected activity because these systems are not yet cost-effective when compared to conventional grid power. According to EPRI, true economic viability for PV in comparison to utility systems will require PV system costs of less than $3000/kW. Most of the current PV systems were installed at three times that cost.

Nonenergy added value. People will pay a premium for PV for several reasons. Utilities can find distributed benefits by siting PV along a distribution feeder, saving the utility line losses. Systems with storage and power conditioning can boost power service reliability and power quality. Rooftop systems increase insulation and protection for the roof. Parking lot systems provide vehicle shade and weather protection.

Grid interconnection. The small PV systems currently in use don't cause significant technical problems for the utility grid. In some situations, the PV installations can benefit distribution companies through lowered line losses, peak reduction and deferral of upgrades to the distribution system.

Reliability. PV modules have proven reliable and durable, working well after many years in service. The two problem areas seen in older systems are inverter unreliability and output intermittency. Inverter problems can be mitigated by using more small inverters and by designing systems with backup inverters. EPRI is also studying to apply PV direct current (DC) immediately to grid-connected loads to avoid inverter problems. Output intermittency problems can be addressed by adding storage. As a general guideline, according to EPRI, PV can be up to 20% or 25% of load capacity without causing any problems.

Location. PV can be used anywhere in the United States. Systems in the Midwest and Northeast can be as beneficial as in the Southwest because the regional variation in sunlight, and therefore in PV energy cost, is smaller than the variation in retail electricity prices.

Strategies for funding. A number of strategies are being seen. Some utilities have started projects to field PV generation to supplement conventional power supplies. More than 90 utilities have introduced green-pricing programs, and in these programs PV is often the leading clean energy source. Some states and municipalities have established incentives for PV systems to promote PV's benefits. By utilizing green-pricing programs and optimizing the multiple values available through PV power, utilities can include PV projects without impacting their profitability.

Teaming. Most of the cases in the EPRI study were cooperative arrangements among utilities, customers, government agencies, and suppliers. The stakeholders recognized that they have a common goal in implementing renewable energy.

Affordability. Even though PV electricity may be more costly on an energy basis, there appears to be a market for it if the systems are affordable to consumers.

Increased production. PV manufacturers are boosting production to supply increasing demand, both domestic and international.

Retail ready systems

AstroPower and Home Depot began offering AstroPower's residential solar electric power systems through select Home Depot stores in San Diego, California in 2001, then expanded the program to include Delaware, New Jersey, and New York in 2002. AstroPower's SunUPS and SunLine home power systems are now featured in Home Depot stores in select areas. Through displays at each participating store, customers learn how easy it is to generate their own clean electricity with AstroPower solar electric home power systems.

The systems are being sold and installed under Home Depot's *At-Home Services* initiative, a full-service program providing product sales, installation, and service. Home Depot also offers two financing options to customers.

> *Expansion of this program reinforces the commitment Home Depot has made to bring the most environmentally friendly energy solutions to its customers. A combination of early success in San Diego and heightened consumer interest in our products on the East Coast prompted us to move quickly forward with the expansion. Together, we're bringing a terrific solution to consumers concerned about rising utility costs and electricity reliability.*
>
> HOWARD WENGER,
> VICE PRESIDENT, ASTROPOWER
> NORTH AMERICAN BUSINESS

The AstroPower residential systems are complete packaged systems generating solar electric power and delivering it through a home's existing circuits. As the systems operate in conjunction with the utility grid, electricity not used in the home is sent back to the electrical grid, spinning the utility meter backward and generating a credit on the homeowner's utility bill. AstroPower's SunUPS systems include batteries to provide uninterrupted power 24 hours a day even during utility outages.

> *Home Depot is putting the value of solar electric power right at its customers' fingertips. The mainstream residential market is AstroPower's fastest-growing market. Mainstream consumers seek the benefits solar power can provide, and they choose this solution when given the option.*
>
> Dr. Allen Barnett,
> president and CEO, AstroPower

Energy giant British Petroleum (BP) is also targeting the residential solar market, particularly in California through its Solar Home Solutions program. The program features solar energy to reduce or eliminate monthly electricity bills, boost energy independence, and help keep the environment clean.

> *Homeowners want reliable energy in the form of a complete solution and are increasingly stating their preference for green, environmentally friendly products. That's exactly what we'll deliver: lower energy bills and control in the homeowner hands through a customized and branded system with a long-term panel warranty, backed by a company that's been around for 100 years.*
>
> John Mogford,
> vice president for renewables, BP Group

BP Solar offers customers financing options, access to California Energy Commission rebates up to 50% of the purchase price, a full planning, installation, and commissioning service, and an exclusive in-home display, enabling customers to track their system's electricity production.

An industry first, BP Solar is reaching homeowners through a campaign of direct mail, newsprint, radio, and television advertising. By taking advantage of the Internet as a part of its outreach, BP Solar offers a Solar

Savings Estimator on its web site. Simply by entering their zip code and utility bill, the on-line Solar Savings Estimator allows homeowners to calculate their potential savings, the payback period, and the environmental benefits of having a system for their home.

The California launch signals a new chapter in developing BP Solar's business through a customer-focused approach.

> *A decade ago, the industry focused on developing new technologies and driving down manufacturing costs. Our products are affordable and generate clean, reliable electricity. We now need to become more customer focused.*
>
> MOGFORD

Case Study: Engineering Firm Saves with Solar

AESE an engineering firm from Van Nuys, California, was searching for a way to cut business costs when it settled on PV peak shaving.

> *We recognized the financial benefits of offsetting over 30% of our building energy needs and enhancing the capital value of the building through the addition of a PV system.*
>
> ART BABCOCK,
> PRINCIPAL PARTNER, AESE

AESE also viewed a solar electric installation project as a way to showcase their engineering and construction expertise in environmentally beneficial ways.

AESE settled on a 73 kW system from Shell Solar. It was a natural way to achieve both economic and corporate goals. Now fully operational, the solar electric system reduces AESE's electricity costs most dramatically when time-of-use rates are highest (1:00 P.M.–5:00 P.M. weekdays).

Often referred to as *peak shaving*, the electricity generated by AESE's solar electric system *smoothes* the peaks off those high rates by reducing the amount of electricity needed from their local utility. For AESE and most businesses this is when power usage is heaviest for lighting, workstations, etc. Fortunately, this is also the time when solar modules generate maximum electrical output, making solar an ideal solution for peak shaving.

System specifications

- system peak capacity: 72.9 kW AC
- total projected system
- electrical output per year: ~150,000 kWh
- PV surface area: 7200 sq ft
- solar modules: 1080 Shell SP75s
- mounting method: open rack
- other system components: energy meter and display

> By working closely with Los Angeles City inspectors and Shell Solar engineers, we successfully addressed permitting and other challenges and, in doing so, created a model for future installations within the City of Los Angeles.
>
> BABCOCK

The most immediate benefit is roughly a 30% reduction in the cost of electricity for the building, which translates into savings of several thousand dollars per month. AESE's installation routinely achieves the rated electrical output, and even sees significant power generation (typically up to 50% of rated output) on cloudy or foggy days.

Considering the incentives received from the Los Angeles Department of Water and Power and Southern California Gas, the system will pay for itself in approximately 4 to 5 years. With an anticipated useful system life in excess of 25 years, AESE can expect 20 years of free electricity. Plus, AESE expects the maintenance or repair costs during that time period to be *virtually zero*.

DOE Solar Buildings Program

The Solar Buildings Program was established by the U.S. DOE's Office of Power Technologies to develop solar technologies that have the potential to provide cost-competitive energy for buildings. Such technologies include solar systems to provide hot water and space heating and cooling for residential and commercial buildings. The program is also investigating innovative technologies, such as solar absorption cooling, which may provide additional savings in the future.

DOE is working with the solar industry and its customers to overcome barriers inhibiting market expansion for solar water heating and space heating and cooling. The program's customer-focused strategy is based on gathering information from home buyers, builders, and utilities to identify their requirements for buying and using solar technology. Application-specific commercialization roadmaps, involving a cooperative effort by DOE, the solar industry, and key customers, are guiding the program in its pursuit of displacing 0.16 quads of energy (164 trillion British thermal units [Btu]) by the year 2020.

The program's solar thermal focus is driven by the need for thermal energy for water heating and space heating and cooling applications. Researchers within the program are working on a range of solar thermal technologies that generate hot water and heated air for residential and commercial use. In addition to further developing established solar thermal technologies to compete with conventional building systems, the program develops innovative, next-generation technologies, such as solar absorption cooling.

Market deployment, quality assurance, research and development, industry partnerships, and solar manufacturing assistance activities are all part of DOE's efforts to stimulate the expanding use of solar technologies in buildings and reduce the country's dependence on fossil fuels.

The Solar Buildings Program is managed through a team consisting of representatives from DOE's Office of Power Technologies Solar Buildings Program, the NREL, and the Solar Energy Industries Association (SEIA).

Research and development

DOE is sponsoring numerous research and development activities under the Solar Buildings Program to improve existing solar buildings technologies, find new applications, and stimulate new technologies. Information received through customer workshops and surveys allows research and development efforts to be targeted on areas deemed critical by technology stakeholders.

Two national laboratories (NREL and Sandia National Laboratories), several universities, and numerous industry partners, work collaboratively on many solar building technologies.

Researchers under the Solar Buildings Program are now investigating:

- using innovative concepts to reduce the cost of solar water-heating systems
- researching new applications for solar water-heating systems
- identifying opportunities for the use of plastics in residential solar water-heating systems
- using solar energy systems, including compound parabolic concentrating collectors, to power air-conditioning systems
- developing a new selective coating for flat-plate collectors

Case Study: Solar Patriot House Displays Green Design

The Solar Patriot has two different PV systems that provide sufficient power for the home to operate completely independent of the utility grid—even on cloudy days and cold winter nights. The house features building-integrated PV (BIPV) shingles and low profile solar panels. (*see* Fig. 2–3)

Fig. 2–3 The Solar Patriot House (photo courtesy of Shell Solar)

In addition, a solar hot water heating system accommodates all of this 3000 square foot, four-bedroom, two-story home's hot water needs.

Together, the PV systems will generate 6 kW of energy; enough to handle the demand for all electricity usage in the home. Electricity is stored in eight

75 amp batteries, and an inverter converts the DC electricity produced by the panels and thin-film BIPV shingles to 120 volt AC current.

According to BP Solar, manufacturer of the solar arrays mounted in the center section of the Solar Patriot's roof, solar cells operate on the principle that electricity will flow between two different semiconductors when they are placed in contact with one another and exposed to light (the semiconductors are made of silicon refined to 99.9999% purity).

To make the most of the power it receives from the sun, the Solar Patriot uses R-21 insulation in walls and R-38 in the ceiling. High performance, double-pane windows with low-e coatings round out a highly efficient shell. Passive solar strategies—those that function without any moving parts—include placement of the majority of windows on the home's south side, with appropriate shading, to take optimum advantage of seasonal variations in the sun's path.

An efficient building exterior is important, because it's the part of the home that keeps the outside out and the inside in. How well the building envelope does its job determines everything from the size (and cost) of the heating, cooling, and ventilation system, to the amount of moisture penetrating the house, the amount of ultraviolet (UV) light entering the home (causing furniture, carpets, and finishes to prematurely fade), and the size of the monthly heating and cooling bills.

Even though it costs a bit more to outfit a home with higher levels of insulation, better quality low-e windows, and other green building products and materials, these strategies usually let homes use a smaller heating ventilating and air conditioning (HVAC) system. The reduced cost of the HVAC system can offset much of the increase in material costs required for a better envelope. Even if well-insulated construction still costs more, the increase generally pays for itself in the first few years' worth of utility bills.

As fuel prices continue to rise, the simple payback period—the time it takes to recoup the difference in construction costs between conventional and high performance materials and equipment—gets shorter.

Energy efficient lighting and appliances round out the overall *whole building design* approach by helping reduce the home's overall energy use by as much as 40%. By itself, compact fluorescent lighting throughout the home will yield a 20% savings in the average homeowner's energy use. Together, the high performance building exterior and energy efficient lighting and appliances will reduce the home's overall construction cost by reducing the size of the solar energy system by more than 60%.

A unique feature of this zero-energy home is the fact that its owner will be able to sell any unused, excess solar energy back to the utility for use in the clean power market. Virginia, along with 31 other states, has passed legislation enabling net metering for individual homeowners who produce their own power.

Here are some of the Solar Patriot's other renewable energy and sustainable design features:

- The solar water heater, produced by Duke Solar, is an evacuated tube, nonimaging solar system that stores hot water in a superinsulated tank.
- The insulation in the house and the double pane low-e windows provide maximum insulation to hold heat inside in winter and air-conditioned air in summer. These features also contribute to occupant comfort by minimizing drafts and assuring interior perimeter surface temperatures that are more in line with the ambient indoor air.
- Highly efficient, DOE Energy Star-rated appliances and systems by manufacturers like Whirlpool, Fisher & Paykel, ECR Technology, and others.
- Energy efficient lighting by Osram Sylvania and Maxlite.
- Solar-electric roofing shingles by Baekert-ECD.
- Traditional solar electric panels by BP Solar.
- Electric interconnection and inverters by Xantrex.

Differentiating solar radiation types

Sunlight arrives at earth in three different ways. They are collectively referred to as global radiation. The three types follow:

- *Direct radiation.* When sunlight travels from the sun to the ground with only a slight scattering of the sun's rays in the atmosphere. At any time of year, about 80% of the Sahara desert's total solar radiation is from direct sunlight.

- *Diffuse radiation.* When sunlight is scattered by clouds or haze. In Northern Europe, the proportion of diffuse light can be up to 80% of the total solar radiation in winter and up to 50% in the summer.

- *Albedo radiation.* When sunlight is reflected from the ground. For example, white surfaces such as snow reflect the sun's rays and stay cold. In contrast, dark surfaces absorb solar energy and become warm.

Solar factoids

Earth receives as much energy from sunlight in 20 days as is believed to be stored in its entire reserves of coal, oil, and natural gas.

Some 2,000 years ago the Greeks used mirrors to focus the sun's rays on Roman ships, causing them to catch fire.

There are three main types of solar power systems: solar buildings, solar thermal concentration systems, and PV cells.

The term PV comes from the Greek *phos*, which means light and *volt*, from the scientist Alessandro Volta. In other words, PV literally means *light-electricity*.

With today's technology, houses with passive solar design and efficient insulation can save as much as 99% of energy used for space heating and cooling.

Although the sun releases 95% of its energy as visible light, it also produces infrared and ultraviolet rays.

Each part of the solar spectrum is associated with a different energy. Within the visible portion of the solar spectrum, for example, red light is at the low-energy end and violet light is at the high-energy end, with 50% more energy than red light.
(Courtesy of DOE and Shell Solar)

Types of Solar Cells

PV conversion, which enables the sun's rays to be converted into electricity, is based in solar cells manufactured from silicon. Silicon is one of the most abundant materials on the planet, in the form of quartz sand, and is an environmentally friendly material.

PV technology was pioneered in 1954, when Bell Lab scientists displayed the first silicon solar cell at the National Academy of Science in Washington, D.C. The cell contained specially treated silicon strips which, when exposed to light, powered an FM radio. The *New York Times* hailed the new technology, proclaiming *Vast Power of the Sun is Tapped by Battery Using Sand Ingredient*.

The silicon solar battery was thought to *mark the beginning of a new era, leading to the realization of one of mankind's most cherished dreams - the harnessing of the almost limitless energy of the sun for the uses of civilization.*

The *Times* may have overstated things a bit, since PV is still a niche technology, due primarily to the expense of the equipment.

A number of laboratories and manufacturers are working hard to improve generation while decreasing costs, which has lead to a number of technologies and approaches to PV.

Solar cells can be made from a variety of semiconductor materials. Also, each cell has an electrical contact with the semiconductor that carries the electrons away from the cell and into the electrical circuit where they can put their energy to use, lighting light bulbs or charging batteries.

The whole cell is encapsulated in a transparent cover providing environmental protection and containing an anti-reflection coating to make sure that as little light as possible is reflected away from the cells.

The different types of solar cells are:

- crystalline silicon
- thin film
- nonsilicon compound thin film
- nano-crystalline
- fullerene

Case Study: Santa Rita Jail Boasts Biggest Roof Installation

Almost three acres of the Santa Rita Jail's rooftop in Dublin, California are covered with PV solar collectors, expanding a previous 640 kW array to 1.18 MW and making the facility the largest rooftop solar installation in the United States.

PowerLight Corp. of Berkeley, California, installed the first solar array on the jail during the summer of 2001. The PV system performed so well that PowerLight was commissioned to increase the county's initial PV system. The combined project has reduced the facility's peak summer demand for grid-generated power by 35%. More than 2.5 million kWh/year are diverted from California's grid by the Santa Rita project. The savings benefit all of the state's consumers by reducing grid power purchases, most of which occur during peak electrical demand hours—at times when the state transmission lines are the most constrained.

Based on local electricity rates, the total project savings for the county total about $425,000 in the first year of operation with $15 million in net savings projected over the 25-year life of the installation.

Partial funding for the projects came from the California Energy Commission's Emerging Renewable Buydown program, incentives from California Public Utility Commission (PUC), and prior energy efficiency incentive payments.

Alameda County also received a low interest rate energy efficiency loan from the energy commission and did not have to authorize any general fund revenues to finance its solar electric generation and energy efficiency projects. The electrical cost savings will pay the debt service for the loan.

PowerLight worked with CMS Viron to integrate energy efficiency measures with the solar generation system to maximize energy savings. PowerLight's PowerGuard solar roof tiles insulate and protect the roof while generating solar power. The project incorporates solar cells from AstroPower and BP Solar.

Case Study: Largest Solar Installation in the United Kingdom

The new TXU Europe building in Ipswich holds the U.K. record for largest PV facility, with 87,000 high performance, grid-connected mono-crystalline silicon PV cells, supplied by BP Solar.

The 200 kW project features cells housed in glass panels to serve the dual purpose of replacing conventional materials while simultaneously harnessing energy from natural light to produce clean energy. The 200 kW output is more than 10% of the maximum power requirement of the building.

BP Solar says the building is a showcase for BIPV technology, widely regarded as a key feature of buildings of the future.

> *The project is significant because it proves that the technologies are available to deliver practical solar energy systems. It is also meeting the government's challenge to the construction industry to integrate renewable energy into buildings, and because we have been involved in the design process from the very beginning of the project, we have been able to develop a design that will maximize the benefits of PV, in terms of performance and substitution of conventional cladding material.*
>
> Ray Noble,
> business executive manager, BP Solar

The £35 million building project, which is beginning to take shape in the heart of the Ipswich Village redevelopment area, is creating a new sustainable clean energy headquarters for TXU-Europe, the United Kingdom's largest domestic electricity supplier.

The system will also cut carbon dioxide (CO_2) emissions. It is estimated that over the course of a year emissions of CO_2 will be reduced by some 140 tons and the PV panels will generate a significant percentage of the site's total electricity.

From an architectural point of view, the aesthetics of the building are stunning. Four types of glass-to-glass PV modules are incorporated into the actual fabric of the six-story building structure, using a standard curtain wall system. There are also more than 600 standard BPS laminates installed as screening around the rooftop plant areas to utilize otherwise *spare* roof space for additional PV generation.

The solar modules are integrated on the south facing facades, spandrel panels, modesty panels, and on the atrium roof.

Grid-connected solar technologies in both residential and commercial buildings are growing at rates in excess of 30% per year and according to a recent survey among local councils in the United Kingdom, it was easily the most favored technology among those councils currently planning a renewable energy project.

New or refurbished schemes can exploit BIPV, which provide the option of PV used as a substitute building material instead of roof tiles or glass walls. BIPV offers both striking design features and displaces the original building material, thereby reducing the cost of introducing PV. Contrary to popular understanding, PV does not require full sun to provide energy. A typical British gray day will also generate sufficient sunlight to produce electricity. The technology is simple for installers to manage and once installed is very low maintenance.

NOBLE

Case Study: 4000 Solar Systems for Rural Telecom

BP Solar Latin America is providing 4000 PV systems for a rural telecommunication project in Peru. The systems will enable remote community centers that currently lack access to the electricity grid to conduct voice and data transmissions, which may include telephone, fax, and internet services.

Financing for the $2.4 million project was arranged through the U.S. Export-Import Bank, with AllFirst bank acting at intermediary.

With this project, known as Fitel, 4000 PV systems (approximately 580 kW) are being provided by BP Solar Latin America. The installation of this rural telecommunication system in Peru will benefit close to 3,000,000 people, approximately 10% of the population, providing them with telecom access and connection.

BP Solar has also installed 1.33 MW of solar PV systems in 1852 schools throughout 11 states in Brazil.

Case Study: 3000 Solar Panels Save University $50,000 Annually

Shell Solar is involved in one of the largest solar electric installations at a public university in the United States. The installation of more than 3000 solar panels at California State University Northridge (CSUN) is expected to save the university more than $50,000 annually in energy costs while at the same time contributing to a cleaner environment, through the use of clean, renewable energy such as solar electricity (Fig. 2–4).

Fig. 2–4 Solar Parking Panels at CSUN
(photo courtesy of Shell Solar)

For the past two decades, Northridge has been very active in seeking new and innovative technologies to reduce its energy bills. This project is a good example of the university's commitment to promoting environmentally friendly technologies, support energy conservation, and reduce its energy costs.

The Shell Solar modules, which are doubling as shading in student parking lot, can generate 75 watts of power each, producing a peak generating capacity of 225 kW. Much of this power will be generated exactly when it is needed most—between 1:00 P.M. and 5:00 P.M. during summer months.

PV cells in the panels absorb the sun's rays, creating DC power that is directed to a substation where it is converted to alternating power. It is then increased to 4160 volts of energy and fed into a power grid that distributes electricity throughout the campus.

In addition to saving energy, the use of the PV cells is easing the campus' impact on the environment. According to the Environmental Protection Agency (EPA), by using 225 kW of PV capacity, the university reduces carbon emissions equal to the amount emitted by an average of 36 passenger cars driving 20,000 miles per year.

The savings realized by the university will allow more funds to be directed to educational and student programs. Also, the project has provided a team of CSUN engineering students an opportunity to put to use the skills they have learned in the classroom.

> *It's probably been the greatest opportunity I've had so far in school. I was able to apply what I've learned in the classroom to a real world situation. In the classroom you deal with numbers and a lot of math, but working on this project I really learned what it takes to get something built and how to deal with all sorts of people.*
>
> JOSH GALLO,
> PROJECT MANAGER, CSUN ELECTRICAL ENGINEERING STUDENT

Case Study: Peru Houses First Latin American Solar Grid-Connect System

Eco Solar, a BP Solar distributor, is working with Luz del Sur, a power distribution company, serving customers in southern Lima to develop up to 500 kW of grid-connected solar power for Luz del Sur customers. The contract, the first of its type in Latin America, has an initial objective of installing 100 small solar systems.

Eco Solar developed a pioneer project in Peru, successfully installing a 2.6 kW pilot grid-connect system in a four-story building in Peru's capital. This system demonstrated how grid connected PV systems work, and it helped to progress the discussion between Eco Solar and the utility company.

An important element of this project is the application of net metering. When the PV system is generating less power than required by the user, the utility provides the difference, spinning the meter forward. When it generates more power than required, the excess electricity flows into the utility grid, spinning the electric meter backwards and building a credit against the utility bill.

The solar electricity from BP Solar provides a high quality feature that actually saves users money, reducing in some cases power costs by 50% to 100%, while also contributing to a cleaner environment.

Another PV grid tie system pilot project is being developed in Brazil, where BP Solar signed a contract with CEPEL Federal Electric Research Center—to supply and install a 16kW PV grid tie system.

Case Study: Fun in the Sun

The Pacific Wheel at Pacific Park is the world's first solar-powered Ferris wheel. An innovative and eye-catching use of solar power, it rises nine stories above the Santa Monica Pier deck. Since its construction in 1998, its revolving lights have become a familiar nighttime sight on the Southern California coast. It's one of the few attractions in the area that is not dependent on conventional electricity (Fig. 2–5).

FIG. 2–5 THE PACIFIC WHEEL IN CALIFORNIA RUNS ON SOLAR POWER
(PHOTO COURTESY OF SHELL SOLAR)

The 130-foot wheel generates more than 71,000 kWh of PV power from the sun's rays. On cloudy days, the wheel relies on conventional power.

Mounted on top of the park's loading area are 650 PV modules. The modules generate electricity to power the wheel.

A bridge builder, George Washington Gale Ferris, invented the Ferris wheel in 1893 for the World's Columbian Exposition in Chicago, commemorating the 400th anniversary of Christopher Columbus' arrival in this hemisphere. It cost an estimated $350,000 and was driven by two 1000 horsepower reversible engines.

Case Study: Learning From Light Brightens Schools

American Electric Power (AEP) is conducting a program called *Learning From Light*, installing PVs in more than 100 schools. The first school installation was at Bluffsview Elementary School in Worthington, Ohio, in 1999. The program passed 100 installations in 2003.

The solar equipment helps get students excited about science and math. Learning from Light is a hands-on program for teaching students about energy resources. It was developed by AEP with the Foundation for Environmental Education and the DOE. The program is available to any educational facility through local partnerships with the greatest participation among elementary and middle schools. Some international schools are also participating in the program. One, in Porvenir, Bolivia, is using its PV system to provide the only electricity in the village.

Teachers at participating schools use a curriculum developed by the foundation and National Energy Education Development project to make classroom connections between the PV installation at the school and in-class activities. Proficiency test scores in math and science improved at Bluffsview following installation of the solar system—the true mission of the program.

A typical school installation includes a 1000-watt, ground-mounted PV panel system with a 10 ft x 8 ft footprint. Panels are usually placed in a high traffic location so students, teachers, and public become familiar with the technology. Once the solar panels are up, students can graph how much power is generated hour by hour, while another graph shows them how much power their school is consuming.

Total generating capacity from all the schools exceeds 100 kW. The DOE Office of Renewable Energy certifies each project under its Million Solar Roofs and Energy$mart Schools programs.

The systems typically cost schools about $10,000. With data collection systems and teacher training, the full system is valued around $12,000. AEP facilities contract arrangements between schools and vendors and help schools apply for grants. Some states offer grants or incentives, local sponsors often contribute to the program and some AEP power plants have adopted schools for the program.

Case Study: Staying Green at Hybrid Parker Ranch

Hawaii's Parker Ranch boasts one of the largest solar/wind power systems in the world. AstroPower, Bergey Wind Corp. and PowerLight Corp. teamed up to build the $2 million, 225 kW hybrid system, which currently powers the water pumping demand of the 91,050 hectare cattle farm. PV equipment and five wind turbines generate approximately 90% of the power needed to provide water to the grazing fields (Fig. 2–6).

FIG. 2–6 PV AND WIND TURBINES AT PARKER RANCH

The hybrid wind/solar concept is by no means new. It was first developed after WWII, but it has only been in the last 20 years that commercial implementation has taken place.

Often when there is a lot of sun, there is little wind, and vice versa, so shortages can happen, but the power generated at the ranch is either stored in batteries or fed into the utility grid and used as needed. With a battery system, a generator can be added to cover periods of high demand or the unlikely event of no sun or wind. A power tracker follows the sun during the day and five wind turbines gather energy during the night (Fig. 2–7).

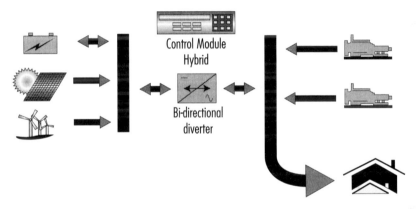

FIG. 2–7 A POWER TRACKER FOLLOWS THE SUN

Parker Ranch selected the system for cost and convenience. The on-site system was more convenient because a power line would have been needed to run 2000 meters (m) up a mountainside. High energy costs in Hawaii also tipped the scales in favor of the renewable technologies. Power prices in Hawaii tend to run as much as 25% higher than in mainland United States.

Equipment

The facility has 175 kW of PV and 50 kW of wind equipment to generate 90% of the power needed to provide drinking water to the Mauna Kea, Mana, and Keamuka grazing fields. The remaining power comes from the local grid. The project is controlled by a supervisory control and data acquisition (SCADA) system, which manages the efficiency of the hybrid system by matching electrical load to available solar and wind energy, which in turn pumps the water up the mountain.

The complex system operates around the clock in a joint effort to ensure a continuous flow of energy. The solar panels gather solar energy with a power tracker rotating with the sun from east to west through the day, while at night when cooler air settles, the five wind turbines gather energy to power a steady flow of electricity. If wind energy is low, a backup generator makes sure the energy supply doesn't falter.

Benefits

The primary benefit of the system is clean, quiet energy. Diesel generators operate most efficiently when running at 80% to 90% of rated capacity, becoming less efficient as the load decreases. This happens because the diesel's combustion chambers don't reach operating temperatures, resulting in carbon buildup on cylinder walls and increased acidity in lubricating oils. This effect can easily double the maintenance requirements of the unit and halve its operating life.

Hybrid systems address the shortcomings of standalone diesel generators by incorporating equipment that optimizes generator performance while more effectively providing power for varying loads.

Future views

The hybrid solar/wind power system is a useful way to generate electricity in remote areas. The advantages of these systems could also encourage people to use them for residential applications. In some parts of the United States, homeowners who have a hybrid system can sell excess power to the grid for extra income. The life span of such machinery can last for decades and there are some installations that have been operational since 1946.

> *I think that in the future, this type of renewable energy will touch everyone's life. Right now it does so in many ways that people don't even see. Examples of this are solar navigation systems, solar emergency call boxes, solar and wind communication systems. This equipment is the future.*
>
> MARIO MAIALE, SUN PRODUCTS

Mass implementation will require proper marketing. Research shows the payback period for hybrid systems tends to run about four years, which can be a selling point. However, the initial cost can scare off investors.

> *Cost is a major issue with the United States and that is where the difference lies with the Europeans. If the Europeans were to install such a system they tend to think more about environmental consequences.*
>
> ANDY KRUSE, SOUTHWEST WINDPOWER

Case Study: HCPV—Catching Rays in Arizona

APS is the proud owner of the world's first commercial high concentration PV (HCPV) solar arrays in Glendale, Arizona. APS contracted Amonix Inc. to build the 100 kW installation, which uses special technology to track the sun's movements and uses special lenses to magnify the sun's rays 250 times onto each solar cell. The HCPV plant supplies enough energy to power more than 160 homes (Figs. 2–8 and 2–9).

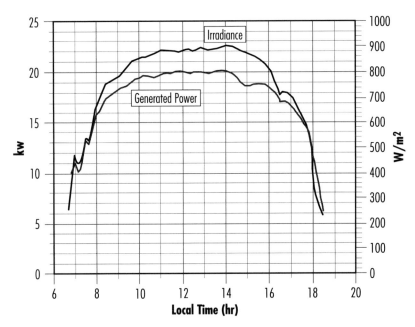

FIG. 2–8 TYPICAL GENERATED POWER FOR THE ARRAY

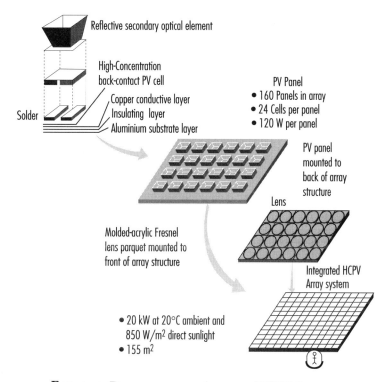

Fig. 2–9 Design of the Amonix HCPV System

The project is based on technology originally developed by EPRI, which based its design on work from Stanford University. Amonix has modified and developed the technology, which is now expected to be the lowest cost PV technology when it is manufactured in volume.

> *It is highly efficient and the plant will be very reliable. The units are also flexible in installation. You just need a place that has good sun. It's modular so there isn't a long lead-time between the start of the plant and the start of electrical production. It can start producing power before all the units are installed.*
>
> Vahan Garboushian, president, Amonix

When power demand increases in the future, another 10 or 100 units, whatever is needed, can expand the systems.

APS has been researching, developing, and using solar power for more than 25 years with a belief in the benefits solar technology can offer its customers. The Glendale installation allows more electricity to be generated from smaller and less expensive PV cells than those used in other solar systems because the lenses concentrate the sun's rays. By concentrating the solar energy, the specialized solar cells operate more efficiently than conventional cells.

In fact, the same amount of electricity can be generated from only 1 square centimeter (cm) of cell area under concentration as from 250 square cm of nonconcentrating cells. The cells are a significant cost factor in any PV systems and the Amonix technology directly affects this cost driver.

APS is offering customers the opportunity to pay $2.64 per month to have 15 kWh of their electricity met by solar power through its APS Solar Partners program. This monthly fee funds plants like the Glendale facility and enables research into developing solar technology. The HCPV installation also received a financial boost from Utility PV Group. The land at Glendale's Municipal Airport was provided by the city through a low-cost lease.

How it works

The system uses a Fresnel lens to concentrate the sun's energy on a silicon cell. Four of these silicon cells are mounted onto a strip of aluminum. The cells are then electrically connected by a copper strip placed on top of the aluminum strip while being isolated from it. The solar cells are connected in series, referred to as a string.

Each solar cell has a secondary optical element placed around the cell to redirect any beam dispersion. Each cell on the strip also has a bypass diode to ensure that a single bad cell in the string will not interrupt the flow of current in the string.

Six strips are mounted on an aluminum plate with a heat sink mounted on the backside of the plate opposite the cell strip. The back plate and heat sink will transfer waste heat to the ambient air, preventing active cooling so none of the generated power is used as parasitic power for cooling.

The Fresnel lens is a circular lens having a 53 cm focus point. Twenty-four of these lenses are manufactured as a single unit of parquet. The parquet lens and cell plates are integrated with a support structure aligning the Fresnel lens with the solar cell, positioning the cells at the focus point of the Fresnel lens, and providing an enclosed space for environmental protection of the solar cells. The 25-cell grids are called MegaModules and they measure 13.4 m by 3.3 m by 60 cm.

Efficiency measures

The modules must be pointed accurately toward the sun for maximum efficiency. They are therefore mounted on a tracking structure, which uses a gearless hydraulic drive to position the array to receive the maximum solar exposure. The electronic controller is completely autonomous but has the capability to be monitored or controlled remotely. The controller has a built-in global positioning system (GPS) signal to obtain universal time, which it uses to calculate the sun's position. As the sun's irradiance increases, the inverter will connect to the gridline and begin producing power automatically.

The array automatically starts operating in the early morning, producing power then the sun's irradiance reaches 300 to 400 W/m^2. Output increases as irradiance increases. The inverter automatically disconnects from the grid where there is cloud cover, but continues to track the sun to reconnect when cloud intensity decreases.

Case Study: Australian Solar Tower Breaking Records

Australia will soon be home to the world's first commercial-scale solar tower plant, which will also be the world's tallest manmade structure and one of the world's largest nonhydroelectric renewable energy generators.

The 200 MW plant is expected to cost $431 million to build. It began construction in 2003, with a projected completion date in 2005 or 2006. The tower will have a ground-level greenhouse area to heat air and force it through a tall, narrow chimney. The flowing air drives turbines to produce power. The tower acts like a magnet in attracting sunrays from the air and from the large body of air underneath the collector system (Fig. 2–10).

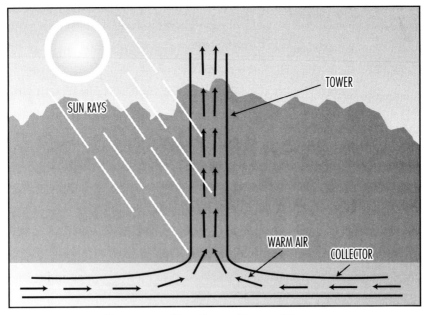

Fig. 2–10 The Solar Power Tower

The first solar tower concept was designed in the early 1980s, but its use has been restricted to small-scale demonstration projects. The Australian plant will, according to its developers, compete with other forms of generation in the nation's power market.

The plant is sited near Buronga in Wentworth Shire, Victoria. It will have a 1000 meter-high tower and a collector system with a radius of 3.5 km. It will produce enough power for about 200,000 homes and will offset the emission of almost 900,000 tons of greenhouse gases annually.

Australia has committed to reducing greenhouse gas emissions through its Mandated Renewable Energy Target, under which 2% of power generated must come from renewable sources by 2010. The power tower will generate 560 gigawatt-hour (GWh) annually, accounting for around 7% of the mandated target. The tower is being developed by Australian company EnviroMission through an agreement with Schlaich Bergermann of Germany, the designers of the technology.

Jorg Schlaich designed the original solar tower technology, and Schlaich Bergermann constructed a 50 kW tower in Spain, which is the prototype for the Australian project.

Financing

Late in 2001, EnviroMission signed a memorandum of understanding with a local utility for a power purchase agreement, which will help EnviroMission with financing issues. The project has also received Major Project Facilitation status from the government, which means investors receive additional incentives, including accelerate depreciation. The plant will also receive renewable credits, which is a revenue stream generated by a levy on all nonrenewable forms of generation. EnviroMission's optimization study, completed in 2002, confirmed the economic viability of the plant.

The project is also expected to bring in revenues from nonenergy sources. These could include viewing platforms on the tower and leasing land for agriculture. EnviroMission hopes to build five solar towers in Australia by 2010. Energen Global, part owner in EnviroMission, plans to launch the technology in the United States through its California office.

Design and construction

The tower is being constructed on Tapio Station, a large grazing area in New South Wales. The site was chosen predominantly for its proximity to the national grid. The site is only 7.5 kilometers (km) from a terminal station, and it will be able to take advantage of premium prices for green electricity in the New South Wales market.

The plant will have a collector system that stands slightly above the ground but rises toward the central tower. The collector will be composed of a transparent material with heat enhancing properties.

The 1000 m tower will be built from reinforced concrete and strengthened by horizontal metal supports that can also act as platforms. The air temperature under the collector roof will be around 30°C and the wind speed will be about 32 km/h.

The sun's rays will heat the large body of air underneath the collector system. As the hot air attempts to rise, it is forced toward the central tower, passing through 32 shrouded wind turbines set horizontally in the transition area. The tower creates an updraft, drawing air inward. The turbines will be purposely built from lightweight alloy materials, with 10 blades coupled to synchronous generators. The turbines will receive air at around 60°C to 70°C.

One innovative aspect of the tower's design is its ability to generate electricity 24 hours a day. It will be most efficient during the day when the sun's rays are more intense, which is also when electricity prices peak.

The use of heat-storing material on the ground underneath the collector will allow the plant to generate power throughout the night as well.

Based on the experience of the now-decommissioned prototype plant in Spain, the new tower is expected to need only minimal maintenance, due to the absence of pressure parts and elevated temperatures. The turbines used are more reliable than standard wind turbines, so overall plant availability is expected to be close to 99%. The collector itself will also need very little maintenance.

Wind Energy

Wind energy, also called wind power, refers to the process by which wind is used to generate mechanical power or electricity. Wind turbines convert the kinetic energy in the wind into mechanical power. This mechanical power can be used for specific tasks—grinding grain or pumping water—or a generator can convert the mechanical power into electricity.

Wind is actually a form of solar energy because the uneven heating of the atmosphere by the sun, the irregularities of the earth's surface, and rotation of the earth cause winds. Wind flow patterns are modified by the earth's terrain, bodies of water, and vegetative cover. This wind flow, or motion energy, is then used by wind turbines to generate electricity.

Since earliest recorded history, man has used wind power for a variety of tasks. Wind power has been used to move ships, grind grain and pump water for centuries.

There is evidence that wind energy was used to propel boats along the Nile River as early as 5000 B.C. Within several centuries before Christ, simple windmills were used in China to pump water.

In the United States, millions of windmills were erected as the American West was developed during the late nineteenth century. Most of them were used to pump water for farms and ranches. By 1900, small electric wind systems were developed to generate DC, but most of these units fell into disuse as inexpensive grid power was extended to rural areas during the 1930s. By 1910, wind turbine generators were producing electricity in many European countries.

Types of turbines

Wind turbines, like aircraft propeller blades, turn in the moving air and power an electric generator that supplies an electric current. Modern wind turbines fall into two basic groups: (1) horizontal-axis variety, like the traditional farm windmills used for pumping water, and (2) vertical-axis design, like the eggbeater-style Darrius model, named after its French inventor.

Modern wind technology takes advantage of advances in materials, engineering, electronics, and aerodynamics. Wind turbines are often grouped together into a single wind power plant, also known as a wind farm, and generate bulk electrical power (Fig. 3–1).

FIG. 3–1 WIND TURBINES CAPTURE THE POWER OF WIND
(PHOTO COURTESY OF NEG MICON)

Electricity from these turbines is fed into the local utility grid and distributed to customers just as it is with conventional power plants.

Today's wind turbines are available in a variety of sizes and power ratings. The largest machines have propellers spanning more than the length of a football field and standing 20 building stories high.

At the other end of the size spectrum, a small home-sized wind machine has rotors between 8 ft and 25 ft in diameter and stands upwards of 30 ft.

All electric-generating wind turbines, no matter what size, are comprised of a few basic components: the rotor (the part that actually spins in the wind), the electrical generator, a speed control system, and a tower. Some wind machines have fail-safe shutdown systems so that if part of the machine fails, the shutdown systems turn the blades out of the wind or puts on brakes.

U.S. wind resources

Wind energy is very abundant in many parts of the United States. Wind resources are characterized by wind-power density classes, ranging from class 1 (the lowest) to class 7 (the highest). Good wind resources (class 3 and above), which have an average annual wind speed of at least 13 miles per hour, are found along the east coast, the Appalachian Mountain chain, the Great Plains, the Pacific Northwest, and some other locations (*see* Table 3–1).

TABLE 3–1
TOP 20 STATES FOR WIND POTENTIAL

1. North Dakota	8. Oklahoma	15. New York
2. Texas	9. Minnesota	16. Illinois
3. Kansas	10. Iowa	17. California
4. South Dakota	11. Colorado	18. Wisconsin
5. Montana	12. New Mexico	19. Maine
6. Nebraska	13. Idaho	20. Missouri
7. Wyoming	14. Michigan	

North Dakota, alone, has enough energy from class 4 and higher winds to supply 36% of the electricity of the lower 48 states. Of course, it would be impractical to move electricity everywhere from North Dakota.

Wind speed is a critical feature of wind resources, because the energy in wind is proportional to the cube of the wind speed. In other words, a stronger wind means a lot more power.

Iowa may leap up the list of wind generating states by 2006 with the expected completion of MidAmerican Energy's 310 MW wind generation facility. The wind farm will be the largest land-based wind project in the world when it is fully on-line. MidAmerican announced the project in 2003 as part of a plan to freeze power rates through 2010. The wind project is expected to cost $323 million. The project will have 180 to 200 wind turbines of 1.5 MW to 1.65 MW. The project is expected to begin generating power by the end of 2004 and to reach full capacity by the end of 2006. Iowa's government has stated a goal for the state to become energy independent and to develop into a national leader in renewable energy.

Falling costs

Although the cost of wind power has decreased dramatically in the past 10 years, the technology still requires a higher initial investment than fossil-fueled generators. Roughly 80% of the cost is the machinery, with the balance being the site preparation and installation. If wind generating systems are compared with fossil-fueled systems on a *life-cycle* cost basis (counting fuel and operating expenses for the life of the generator), wind costs are much more competitive with other generating technologies because there is no fuel to purchase and minimal operating expenses.

The major challenge for wind as a source of power is that it is intermittent and does not always blow when electricity is needed. Wind cannot be stored (unless batteries are used), and not all winds can be harnessed to meet the timing of electricity demands.

Good wind sites are often remote locations far from areas of electric power demand. Finally, wind resource development may compete with other uses for the land and those alternative uses may be more highly valued than electricity generation. However, wind turbines can be located on land that is also used for grazing or even farming.

New, utility-scale, wind projects are being built all around the United States today with energy costs ranging from $.039 cents/kWh to $.05 cents/kWh or more. These costs are competitive with the direct operating costs of many conventional forms of electricity generation now—and wind prices are expected to drop even further over the next 10 years.

Since wind is an intermittent electricity generator and does not provide power on an *as needed* basis, it has to compare favorably with the costs saved on fuel from fossil generators.

The wind energy industry has grown steadily over the last 10 years and American companies are now competing aggressively in energy markets across the nation and around the world. The industry, in partnership with the U.S. DOE, continues to expand and develop a full range of highly reliable, efficient wind turbines.

When installed, these new-generation turbines perform at 98% reliability in the field, representing remarkable progress since the technology was first introduced in the early 1980s.

The U.S. DOE is backing the technology through its *Wind Powering America* initiative, which includes goals to power at least 5% of the nation's electricity with wind by 2020, increase the number of states with more than 20 MW of wind to 16 by 2005 and 24 by 2010, and increase federal use of wind energy to 5% by 2010.

One crucial element in the spread of wind energy is the wind energy production tax credit, which makes wind energy more competitive economically. The credit has been in effect for several years now. At this writing it is scheduled to expire at the close of 2003, but wind organizations are pushing for a five-year extension. The credit has expired and been renewed in the past.

Industry statistics

The wind energy industry is growing at about 30% annually and industry experts predict that capacity will jump from the current 31,000 MW to around 80,000 MW in 2006. Europe is the biggest market for wind energy products. Between 2001 and 2006, the European wind market is expected to jump from 23,000 MW to more than 55,000 MW (*see* Table 3–2).

TABLE 3–2
TOP FIVE WIND MARKETS

Country	2002 Additions	2003 Totals
Germany	3,247 MW	12,001 MW
Spain	1,493 MW	4,830 MW
United States	410 MW	4,685 MW
Denmark	497 MW	2,880 MW
India	195 MW	1,702 MW
Source: AWEA and EWEA.		

Global wind power capacity quadrupled from 7600 MW in 1997 to 31,128 MW at the close of 2002. Wind is currently the world's fastest-growing energy source on a percentage basis, with installed generating capacity jumping by an average 32% annually.

New equipment worth $7 billion was brought on-line in 2002 alone, primarily in EU member countries. The United States, accounting for 15% of the global installed wind capacity total, saw a market lull in 2002 as developers awaited extension of the wind energy production tax credit.

The countries of the EU are the biggest buyers of wind technology. In 2002 alone, the EU countries installed 5871 MW of wind capacity. Germany, Spain, and Denmark accounted for 89% of the wind capacity installed in Europe (*see* Table 3–3).

Table 3-3
Global Wind Energy Generating Capacity by Country

Country	2002 Additions	Capacity (MW)
USA	410	4,685
Canada	40	238
North American Total	**450**	**4,923**
Germany	3,247	12,001
Spain	1,493	4,830
Denmark	497	2,880
Italy	103	785
Netherlands	217	688
United Kingdom	87	552
Sweden	35	328
Greece	4	276
Portugal	63	194
France	52	145
Austria	45	139
Ireland	13	137
Belgium	12	44
Finland	2	41
Luxembourg	1	16
EU Total	**5,871**	**23,056**
Norway	80	97
Ukraine	3	44
Poland	5	27
Latvia	22	24
Turkey	0	19
Czech Republic	0.2	7
Russia	0	7
Switzerland	0	5
Hungary	1	2
Estonia	1	2
Romania	0	1
Other Europe	**112**	**235**
India	195	1,702
China	68	468
Japan	140	415
Australia	32	104
Others	NA	225 (est.)
Other Total	**435**	**2,914**
World Total	**6,868**	**31,128**
Source: AWEA and EWEA		

Germany long since established itself as the continent's wind leader. Germany boasted 3247 MW of new installations in 2002, bringing the country's wind capacity to 12,001 MW, fully 4.5% of its power needs. The wind industry employees 45,000 people in Germany and most of the country's turbines are manufactured there. The wind power generation is primarily in the northwestern regions of the country.

Spain installed 1493 MW of capacity in 2002, taking second place in global wind capacity from the United States by reaching installed capacity of 4830 MW. Spain's wind capacity has taken off over the past decade. In 1993, the country had only 52 MW installed. Wind's popularity in Spain began when the government created an incentive for wind power, similar to the incentive that kick-started the German wind market. Spain has large wind installation in Galicia, Aragon, Navarra, and Castilla.

Denmark's 2,880 MW of wind capacity is sufficient to meet 20% of the country's needs. Denmark, which is about the size of the state of Maine in the United States, is the nation that gets the highest percentage of its power from wind. Denmark is also a major manufacturer of wind turbines and other wind energy technology, accounting for approximately half of the turbines being installed worldwide.

The United Kingdom only installed 78 MW of wind capacity in 2002, but the country gave planning permission to another 525, including two offshore wind farms, indicating that the wind market in the United Kingdom is picking up. In 2003, the U.K. government issued a white paper stating that the country plans to turn to renewables rather than nuclear energy to curb emissions, another indicator for growth in the wind market. Germany, Spain, and Denmark accounted for almost 90% of the wind power capacity installed in 2002.

In the United States, 27 states have wind capacity online, and the renewal of the production tax credit (a federal tax incentive for wind generation) bodes well for the future of wind generation in North America. The wind industry has been lobbying for a multiyear extension of the tax credit to stabilize the market.

> *Wind energy grew by 10% in the United States in 2002, a good performance given the poor state of the U.S. energy market and the stop-and-go-and-stop-again policy signals directed at our industry. With steady supportive policies, wind power could grow at a sustained pace closer to that of Europe, and provide well over 6% of U.S. electricity by 2020.*
>
> RANDALL SWISHER, EXECUTIVE DIRECTOR
> AMERICAN WIND ENERGY ASSOCIATION (AWEA)

The AWEA believes that the current growth trends suggest that by 2020, wind could easily account for 6% of U.S. generation with 100,000 MW of installed capacity. Wind is currently only 0.3% of U.S. generation. Figure 3–2 shows wind turbines sharing a cornfield.

FIG. 3–2 WIND ENERGY IS A GROWING PART OF THE ENERGY MIX IN THE UNITED STATES

One recent boost to the U.S. wind market was the purchase of some Enron Wind Systems assets by the General Electric (GE) Corporation in 2002. GE formed a new division, GE Wind Energy, and committed itself to the technology following that acquisition. GE is bringing its global brand name, energy expertise, marketing machine, and investment capital to the wind industry. The company sees synergies with other divisions, including plastics, transportation, and power control.

The GE wind division has 900 kW and 1.5 MW systems available and is working with a prototype 3.2 MW system.

> *We are working to drive further improvements and cost savings for our wind turbine products.*
>
> STEVEN ZWOLINSKI,
> HEAD OF GE'S WIND ENERGY

Energy major Shell also entered the U.S. wind market in 2002 with Shell Wind Energy. Shell bought an 80 MW wind plant in Texas and a 41 MW facility in California. It also has a 50 MW project in Wyoming and jumped into the development or operation of another 1000 MW of wind power in the United States and Europe.

The European Wind Energy Association (EWEA) claims wind power can produce 10% of worldwide energy supply by 2020, even if electricity consumption increases substantially. Denmark and Germany's Schleswig-Holstein are already approaching this 10% figure.

Christophe Bourillon, EWEA executive director, attributes the surge in wind power's popularity to concern about climate change, worries about fossil fuel supplies, and the need to sustain an ever-increasing population.

> *Our association has set targets for Europe alone of 40,000 MW of wind capacity by the year 2010 and 100,000 MW by the year 2020. Wind energy can reduce the amount of greenhouse gases released into the atmosphere, preserve valuable fossil fuel reserves for specialized uses, and help poorer rural countries develop without resorting to polluting technology.*
>
> CHRISTOPHE BOURILLON,
> EXECUTIVE DIRECTOR, EWEA

> *Although there are uncertainties because of the changeable policy environment, we are projecting more than 5000 MW of new growth in the United States over the next decade.*
>
> RANDALL SWISHER,
> EXECUTIVE DIRECTOR, AWEA

Merchant potential

Enron Wind Corp. built the first merchant wind farm, Green Power I, near Palm Springs, Calif. The 22-turbine, 16.5 MW project was built solely to supply emerging green-power markets.

Green Power I began producing power in June 1999. Traditionally, wind power had been sold only under long-term contract to utilities, however, the Green Power I facility was built without contract to sell power through retail marketers.

The project uses advanced Zond Z-750kW Series wind turbines. With 158 and 164-foot rotor diameters, approximately the size of the wingspan of a MD-11 jumbo jet, the Z-750kW wind turbines are the largest manufactured in the United States.

New wind technologies

The U.S. DOE has been working with the nation's wind turbine industry to improve technology and lower costs since 1992. In 1994, DOE announced a $40 million program to develop a new generation of innovative utility wind turbines. The cost-effective turbines are expected to expand markets for U.S. companies in both the United States and in Europe, where competition for new wind projects is driving down costs.

Siting hurdles

Despite its status as the fastest growing renewable energy source, wind power faces numerous obstacles in siting and permitting. This is true for both large projects being built to sell power to utilities and small projects being built for a single user. Regulations and laws governing power project siting are becoming ever more complex, and state and federal siting agencies are not as likely to approve power projects without extensive review. Various interest groups have become more involved in siting procedures as well.

Large wind projects raise many of the same siting issues as other energy projects. There may be concern about truck traffic during construction, health effects of electromagnetic fields from transmission lines, and social issues. Wind projects also face some unique challenges that require special consideration.

Unique considerations

Visual and noise impacts must be addressed. Wind turbines are highly visible structures and often need to be sited in conspicuous locations, such as on ridges or hillsides. They also generate noise that can be bothersome to area residents. Those problems can be mitigated through noise abatement, design adjustments, and other measures. Recent design improvements have greatly decreased the noise generated by today's wind turbines.

Impacts on birds and other local wildlife must also be considered. In some locations, wind turbines and their ancillary equipment have killed raptors, such as hawks and eagles. Pre- and post-construction studies may be necessary to measure the project's impact on wildlife and to create strategies for offsetting them. Soil erosion is another potential problem that may be addressed during the siting process.

Landowner's rights may also be an issue in wind project siting. Wind plants often pay rents or royalties for land use, which can be a benefit to landowners, but it may also raise concerns. A turbine on one resident's land may interfere with a neighbor's ability to develop a wind project.

Guidelines

Successful wind project siting depends on negotiation to balance concerns and benefits. Details vary widely from site to site, but the National Wind Coordinating Committee (NWCC) recommends a few guidelines, including significant public involvement, reasonable time frames, clear decision criteria, coordinated siting processes, expedited judicial review and advance site planning (*see* Table 3–4).

TABLE 3–4
PRINCIPLES OF SUCCESS IN WIND SITING

The NWCC notes eight key elements for the development of a successful process for permitting wind energy facilities:

1. Significant public involvement
2. Issue-oriented process
3. Clear decision criteria
4. Coordinated permitting process
5. Reasonable time frames
6. Advance planning
7. Efficient administrative and judicial review
8. Active compliance monitoring

Source: NWCC

Early involvement allows the public to have its interests factored early in the siting process. Without this, there is a much greater likelihood of later opposition and costly litigation. The public, particularly residents living near a proposed site, should be notified of the siting application, and the siting agency should hold public meetings and accept public comments.

Open siting processes with long delays are a legitimate concern for wind developments. Establishing reasonable time frames for review of applications, hearings, and a final decision from the siting agency is one way to avoid unnecessary delays.

The siting agency should make the criteria for its decisions clear at the beginning. The agency should list all the factors to be considered, specify how the factors are weighed against each other, and set minimum requirements to be met by the project. The factors will vary depending on the circumstances (*see* Table 3–5).

TABLE 3–5
TOP 10 PERMITTING ISSUES

1. Land use
2. Noise
3. Birds and other biological resources
4. Visual resources
5. Soil erosion and water quality
6. Public health and safety
7. Cultural and paleontological resources
8. Socioeconomics, public services, and infrastructure
9. Solid and hazardous wastes
10. Air quality and climate

Source: NWCC

Land use

Unlike most power plants, wind generation projects are not land intensive. On a MW output basis, the land required for a wind project exceeds the amount of land required for most other energy technology, but the physical project footprint covers only a small portion of that land. For example, a 50 MW wind facility may occupy a 1500-acre site, but it will only use 3% to 5% of the total acreage, leaving the remainder available for other uses.

Because wind generation is limited to areas with strong and fairly consistent wind resources, most wind generation is sited in rural and relatively open areas that are often already used for agriculture, grazing, recreation, forest management, or seasonal flood storage.

To ensure that a wind project is compatible with existing land uses, the layout and design of the wind project can be adjusted in a variety of ways, including:

- selecting equipment with minimal guy wires
- placing electrical collection lines underground
- placing maintenance facilities off site
- consolidating equipment on the turbine tower or foundation
- consolidating structures within a selected area
- using the most efficient or largest turbines to minimize the number of turbines required
- increasing turbine spacing to reduce density of machines
- using roadless construction and maintenance techniques
- using existing access roads

Other land use strategies include buffer zones and setbacks to separate wind projects from sensitive or incompatible land uses. Land use agencies in California have established setbacks ranging from two to four times the height of a turbine or a minimum of 500 to 1200 ft from any residential area. Minnesota has established minimum setbacks of 500 ft from occupied dwellings.

For the birds

The problem of birds, especially raptors, flying into wind turbines has been the most controversial, biological consideration affecting wind siting. Wind developments have produced enough bird collisions and deaths to raise concern from wildlife agencies and conservation groups. On the other hand, some large wind facilities have been operating for years with only minor impacts on birds. Smokestacks and radio and television towers have actually been associated with much larger numbers of bird deaths than wind facilities have, and highways and pollution account for a great many as well.

Whether or not this becomes a serious siting issue tends to depend on the protective status or number of bird species involved. Most raptors are protected by state and federal laws and any threat to them may cause siting concerns.

Both the wind industry and government agencies are sponsoring or conducting research into this problem. Studies are under way comparing mortality at lattice and tubular towers and investigating birds' sensory physiology and how it affects their ability to detect components of wind turbines. One study is painting colors on turbine blades to observe birds' reactions.

Wind farms are thought to affect wildlife in several other ways, including:

- direct loss of habitat
- indirect habitat loss from increased human presence, noise, or motion of operating turbines
- habitat alteration resulting from soil erosion or construction of obstacles to migration
- collision with structures, turbine blades, or wires
- electrocution from contact with live electrical wires

The NWCC recommends several strategies for dealing with biological resource siting issues—consultation, surveys, and risk reduction.

Planning and coordination with permitting agencies can reduce the chances of project delays. Most permitting agencies recommend wind developers consult with them and appropriate natural resource protection agencies early in the site selection process to determine the potential for conflicts. It is important to find out whether protected plants and animals inhabit, use, or migrate through the area. Unique or rare habitat types, such as savannas, can raise interest and alternative sites may be needed.

Biological surveys can be helpful, but the timing is important. Some necessary information can only be obtained at a certain time of the year. Protected plants may only bloom for a few weeks or months at a time, and bird use or migration patterns may need to be studied over several seasons or years.

Equipment selection can reduce the risk of high bird mortality, but the best plan is to avoid sites near major bird feeding, roosting, or resting areas. Research is ongoing, but to date there are no designs or modifications that have been statistically proven to significantly reduce the risk of bird collisions. Unless protected plants or animals are involved, most permitting agencies tend to find the noncollision effects of wind development on wildlife to be insignificant.

Sound concerns

Wind generating facilities are very quiet when compared to other types of industrial facilities, but noise is a concern in wind siting because the plants are generally located in rural or low-density residential areas, which are more sensitive to sound issues. These areas also have low background noise levels.

Anything with moving parts makes some sound, including wind turbines, but modern turbine designs are very quiet. In fact, a wind turbine creates much less noise than road traffic, trains, airplanes, or construction activities.

In the past, noise was a problem for the wind industry. Some turbines built in the early 1980s were very noisy and annoying from as far as a mile away. The industry addressed the problem and the machines today are far quieter. The noise from a wind farm, experienced from 750 to 1000 ft away, is no louder than a kitchen refrigerator, according to data from the AWEA (*see* Table 3–6).

TABLE 3–6
ACTIVITY AND NOISE LEVEL COMPARISONS

Source	Noise Level decibel (dB)
Rural night background	20–40
Quiet bedroom	35
Wind farm at 350 m	35–45
Car at 40 mph at 100 m	55
Busy general office	60
Pneumatic drill at 7 m	95
Jet at 250 m	105
Threshold of pain	140

Source: British Wind Energy Association

Today's wind technology is so quiet that you can stand directly beneath a utility-scale turbine and carry on a conversation without raising your voice.

Improvements to design include turning rotors upwind to eliminate the thumping sound made by older units, streamlining towers and nacelles to reduce wind noise and vibration, increasing soundproofing in the nacelle, increasing the efficiency of wind turbine blades, and specially designing gearboxes with sound-dampening buffer pads and flex in the gear wheels to cut mechanical noise.

Small turbines can be noisier for their size than the large machines because the small blades have a higher rotational speed and because less research money has been invested in quieting the small machines.

Visionary planning

Visual or aesthetic concerns are also a common issue in wind siting. Wind projects tend to be located in rural or remote areas with few area residential developments. Potential for visual impact is sometimes considered as part of the evaluation of land use, and the degree to which the visual quality of a project is addressed will vary. Elements that can influence the visual impact of a wind project include spacing, design and uniformity of the turbines, markings on turbines and other structures, spacing of turbines, design and uniformity of the turbines, roads built on slopes, and service buildings.

There is considerable motion in turbine blades and this motion is intensified when the turbines are placed close together, are of different designs, or rotate in different directions. Adequate spacing between turbines and between rows or tiers of turbines mitigates visual impact.

When turbines are sited on ridgelines, the units are visible for greater distances. Against the sloping terrain, surfaces exposed by construction of access roads and turbine pads may contrast with existing soils and vegetation. From a distance, the visual impact of the roads may be greater than that of the turbines. Constructing roads on ridges also may increase erosion.

It is generally recommended that developers contact any agencies with jurisdiction for any maps, plans, guidelines or design standards in that particular area. Design strategies can be used to reduce the visual impact, including:

- using the local landscape to minimize visibility of access and service roads, and to protect soils from erosion
- consolidation of roads or use of grating over vegetation for temporary access without road construction
- use of low-profile building designs
- use of uniform color, structure types, and surface finishes
- consolidating electrical lines and roads into a single right-of-way or corridor
- limiting the size, color, and number of labels on turbines
- limiting size and number of advertising signs on fences and facilities
- using airlift for transport of turbine components and installation

Case Study: Big Spring Keeps on Turning

As early as 1993, TXU Electric and Gas investigated the level of demand for renewable energy in Texas. Encouraged by the enthusiasm of its customers toward green energy, TXU unveiled plans for the $40 million Big Spring wind power project near Midland in December 1998. Developed by York Research, the project has 46 turbines with a total capacity of 34 MW (Fig. 3–3). The final phase, completed in April 1999, saw the commissioning of the largest commercial wind turbines in the world—four Vesta V66 turbines standing approximately 260 ft tall above the elevated plateau of west Texas ranch land.

Fig. 3–3 TXU's Big Spring Wind Project

TXU believes that the project is testament to the fact that as power technologies advance, electricity generated by renewable resources will become more common and economic.

The Big Spring project is built on mesas, rising 195 to 295 ft above the surrounding areas. The winds accelerate as the move up over these mesas. Annual average hub-height wind speeds range from 18.4 to 22.2 mph over the site.

There are three phases to the site:

- Phase I—16 Vesta V47 660 kW turbines
- Phase II—26 Vesta V47s
- Phase III—4 Vesta V66 1650 kW turbines

Projected annual electricity generation for Big Spring is 117 million kWh.

Both turbine models use three rotor blades of epoxy and fiberglass composite. Crosswind separation of the machines is nominally 3.5 rotor diameters. Row-to-row spacing of the machines exceeds 10 rotor diameters to minimize the impact of turbulence from adjacent rotors.

The turbine control system monitors turbine starts and stops under normal operating conditions and also protects the turbines under extreme emergency conditions such as faults caused by a loss of grid load while under power or a component failure. In addition, the system manages the power output of each turbine by pitching the blades and changing the generator slip to maximize energy production while minimizing loads at wind speeds greater than 31 mph.

The control system is operated by a digital computer using Vesta-developed programs. Portions of the system are located in the base of the tower and in the nacelle of the wind turbine. These are linked by fiber-optic lines to minimize interference and damage from lightning.

A key feature of the control system is OptiSlip, which controls loads and spikes from the turbines under high wind speeds. OptiSlip allows the turbine to operate in a similar way to a variable speed machine, preventing the driveline of the machine from experiencing torque spikes.

Offshore aesthetics

When it comes to wind energy, beauty is truly in the eye of the beholder. In the United States some of the opposition to wind installations is due to their appearance, which some find unappealing. The debate over acceptable appearances rose to the forefront recently when Cape Wind Associates planned a wind farm six miles off Cape Cod. The farm would have 170 wind turbines and a total generating capacity of 420 MW (Fig. 3–4).

FIG. 3–4 THE BLYTHE OFFSHORE WIND PROJECT WHILE UNDER CONSTRUCTION (SOURCE: AMEC)

A citizens group moved for an injunction to block the development. The Cape Wind plan calls for wind turbine structures that are 250 ft high. Opposition worries to the project were that the wind farm would be ugly and depress real estate values in an area that holds some of the most expensive real estate in the nation.

The final outcome for the Cape Wind project will be a strong indicator for the future of offshore wind in the United States. Wind developers are watching the battles in Nantucket with interest, and if the project can battle through the many regulatory and NIMBY barriers, there doubtless will be a number of other offshore projects announced.

Canadian Institute for Business and the Environment estimates that the Cape project could replace enough fossil-fired generation annually to cut 4600 tons of SO_2, 120 tons of CO, and 1566 tons of NO_2, which would reduce greenhouse emissions by more than one million tons.

European wind turbine engineers have spoken out for the U.S. project.

Our windmills have become a tourist attraction.
Ferries sell seats so people can come see the windmills.

HANS CHRISTIAN SOERENSEN,
ENGINEER FOR A 20-TURBINE WIND FARM
OFF OF DENMARK'S COAST.

The plant has an arc-shaped footprint less than three miles out from Copenhagen harbor. The 40 MW site has been likened to a piece of kinetic environmental art (*see* Table 3–7).

TABLE 3-7
LARGEST OFFSHORE WIND INSTALLATIONS

Site	Country	On-line	Capacity (MW)
Vindeby	Denmark	1991	4.95
Lely	Holland	1994	2.0
Tuno Knob	Denmark	1995	5.0
Dronten	Holland	1996	11.4
Gotland	Sweden	1997	2.5
Blyth Offshore	United Kingdom	2000	3.8
Middelgrunden	Denmark	2001	40
Uttgrunden	Sweden	2001	10.5
Yttre Stengrund	Sweden	2001	10
Horns Rev	Denmark	2002	80
Large Planned Offshore Projects			
Scheldt River	Holland		100
Ijmuiden	Holland		100
Laeso	Denmark		150
Omo Stalgrunde	Denmark		150
Gedset Rev	Denmark		15
Rodsand	Denmark		600
Lillgrund Bank	Sweden		48
Barsebank	Sweden		750
Kish Bank	Ireland		250+
Arklow Bank	Ireland		500

Source: British Wind Energy Association

In Ireland, Sure Engineer of Dublin is building a project with 200 turbines that will boast a 1000 MW capacity.

> *It really is a fear of the unknown. That's what Americans are dealing with.*
>
> DAN HANNEVIG, SURE ENGINEER

Offshore wind farms are all the rage in Europe. Denmark, Holland, Sweden, and the U.K. have large offshore farms up and running, with many more in the pipeline. Ireland, Belgium, and Germany have projects under development. Denmark already obtains 10% of its electricity from wind, and the government is planning to use offshore wind to meet 40% of demand by 2030.

Ireland recently approved plans for the world's largest offshore wind farm. It will have 200 turbines standing on a sandbank in the Irish Sea producing 520 MW. That's 10% of the country's power demand.

Offshore wind makes a lot of sense, particularly when you consider that wind speeds offshore are much higher on average than those over land. Also, there is less wind shear at sea, which means the turbines receive better quality (less turbulent) wind than their onshore brethren. There is also the huge amount of sea surface to consider—plenty of wide-open space for siting.

The main drawback for offshore wind is the higher cost of construction. Foundations are more expensive, and undersea cables must be run to connect the farm to the shore. In offshore projects, the turbine accounts for only about 45% of the initial capital cost. On land, the turbine is 70% of the project cost (Fig. 3–5).

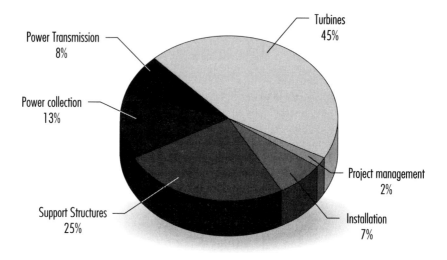

FIG. 3–5 OFFSHORE WIND CAPITAL COSTS

Generally speaking, offshore wind projects must be several hundred MW to a GW in capacity to be economically viable. Therefore, interest in offshore siting has escalated dramatically since the introduction of 1 MW and larger turbines, which make large projects more feasible.

GE Wind Energy's 1.5 MW turbines were the first megawatt-class machines used for offshore generation at the Utgrunden wind farm near Sweden. Two MW Vestas are being used at Horns Rev near Copenhagen. A recent project, the Gunfleet Sands near England, will be the first to use the new GE 3.6 MW machines. It will have 30 of the huge turbines. Turbine sizes are continuing to escalate, with some manufacturers testing models that are 5 MW or larger.

Capacity and output

Wind turbines are most commonly classified by their rated power (RP) at a certain rated wind speed, but annual energy output is actually a more important measure for evaluating a wind turbine's value at a given site. The amount of time a wind turbine produces a given power output is just as important as the level of power output itself.

Wind turbine operators don't get paid for producing a large amount of power for a few minutes (except in rare circumstances). They get paid by the number of kWh their turbines produce in a given time period.

According to the AWEA, the best crude indication of a wind turbine's energy production capabilities is its rotor diameter, which determines its swept area, also called the capture area. A wind turbine may have an impressive RP of 100 kW, but if its rotor diameter is so small that it can't capture this power until the wind speed reaches 40 mph (18 m/s), the wind turbine won't rack up enough time at high power output to produce a reasonable annual energy output.

Expected energy output per year can be reliably calculated when the wind turbine's capacity factor at a given average annual wind speed is known. The capacity factor is simply the wind turbine's actual energy output for the year divided by the energy output if the machine operated at its RP output for the entire year. A reasonable capacity factor would be 0.25 to 0.30. A very good capacity factor would be 0.40.

Capacity factor is very sensitive to the average wind speed. When using the capacity factor to calculate estimated annual energy output, it is extremely important to know the factor at the average wind speed of the intended site. Lacking a calculated capacity factor, the machine's power curve can actually provide a crude indication of the annual energy output of any wind turbine. Using the power curve, one can find the predicted power output at the average wind speed at the wind turbine site.

By calculating the percentage of the RP produced at the average wind speed, one can arrive at a rough capacity factor (RCF) for the wind turbine at that site. By multiplying the RP output by the RCF by the number of hours in a year (8760), a very crude annual energy production can be estimated.

Energy output is also greatly influenced by more subtle features of a wind turbine's design, including the following:

- *Cut-in speed:* The wind speed at which the turbine begins to produce power.

- *Blade shape:* The power a turbine produces at moderate wind speeds, determined largely by blade airfoil shape and geometry.

- *Cutout speed:* The wind speed at which the turbine may be shut down to protect the rotor and drive train machinery from damage.

- *Operating characteristics:* These include low speed, on-off cycling, shutdown behavior, and overall reliability, which together determine the turbine's availability to produce power when the wind speeds are in its operating range.

- *Efficiency:* Especially of drive train components, such as the generator and gearbox.

These features should not be underestimated when looking for ways to improve energy output. In recent years, the U.S. wind industry has begun using seemingly insignificant refinements in blade airfoil shapes to increase annual energy output from 10% to well more than 25%. These increases have helped to dramatically lower the cost of wind-generated energy and increase the number of areas in the United States at which wind plants are feasible.

Transmission issues

By all accounts, the U.S. transmission grid is badly in need of expansions and repair, but transmission upgrades are costly and difficult to arrange. The transmission battles ongoing throughout the country are, or should be, of great interest to the wind energy community. Because many of the country's best sites for wind energy generation are remote, transmission can be a huge siting hurdle.

> We must resolve congestion and make it easier to interconnect new generators of all types to the electric power grid. Right now you have to fight about 140 different wars with 140 different utilities that have their own little idiosyncratic rules about how you can hook up. A standard interconnection procedure from small to big must be adopted across the country. We did it in Texas without much problem and people came there and built power plants, primarily gas and wind.
>
> PAT WOOD III, CHAIRMAN,
> FERC (FORMER CHAIRMAN OF TEXAS PUC)

The current grid was built for monopoly utilities, and is not set up to accept numerous small inputs of power from small generation sites and distributed generation. A variety of groups are working to define the safe, economical way to add small generators to the grid, including the Institute of Electrical and Electronic Engineers (IEEE) and various regulatory bodies. Hopefully, as progress is made on this front, more transmission doors will be opened to wind and other remote sited generation.

The AWEA is fighting for progress on federal policies to ensure that power from new, renewable sources like wind has a fair chance to compete in the nation's electricity markets. The wind industry trade group estimates that several thousand MW of *new* transmission capacity could be obtained in the Midwest alone without building new transmission lines by implementing reforms.

AWEA is calling for positive transmission policies in the Standard Market Design, which FERC is working on.

> *The utility transmission network is the Interstate highway system our electricity generating companies must use to haul their product to market in major population centers. That being the case, it's absolutely critical that electric generators be able to gain access to the transmission network on fair terms along with everyone else, and that the transmission network be big enough to do the job. Lack of transmission capacity is already holding back the development of significant amounts of wind power in the Dakotas.*
>
> RANDALL SWISHER,
> EXECUTIVE DIRECTOR, AWEA

Major transmission issues under consideration include the following:

Rates. The first major issue to be considered is the charge that a wind generator must pay to use the transmission system. Most of the rates in use across the country were not designed with *nontraditional* resources like wind in mind, and they impose discriminatory penalties that can be huge, as much as doubling the wholesale cost of wind-generated electricity. FERC is overhauling the system for setting rates to improve the efficiency of wholesale electricity markets. Its new approach, called the Standard Market Design, should treat wind energy fairly, but there remain many layers of details to iron out for it to work successfully.

Space availability. The allocation of *space* among competing users when portions of the network become congested, much like a freeway during rush hour, is the second issue to consider. Today's congestion management rules are inefficient, letting capacity that could be used go to waste while effectively excluding new entrants like wind energy projects. FERC is proposing a new congestion management system that would significantly expand the capacity of the system to serve all users, including wind.

Process. The third major issue to consider in the building of new transmission lines is the process—planning, permitting, and financing. Many organizations including FERC, the National Governors Association, the DOE, and most industry stakeholders agree the current process for getting new lines planned and built is broken, and the reliability of the nation's electricity supply could be in jeopardy. Certainly, the pace of wind energy development will be drastically slowed—especially in the nation's heartland of the Midwest and interior West—if the process is not fixed and fails to include *high-wind scenarios* that benefit consumers and increase system reliability.

Case Study: Winds of Change in Italy

Italy has little in the way of native fossil fuels, so it must depend heavily on imported fuel, which drives up power generation costs. Recently, wind energy development has been taking advantage of the situation with several farms under construction in southern Italy (Fig. 3–6).

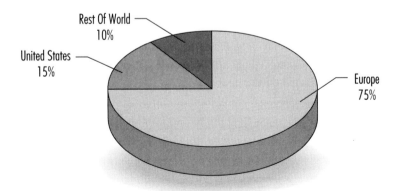

FIG. 3–6 GEOGRAPHIC WIND GENERATION (SOURCE: AWEA)

The government has been in favor of the development of domestically available resources, especially renewables, for diversification, security, and to improve the trade balance. The Italian market also welcomed some hydroelectric and a few geothermal plants, but demand is growing as is the population, and hydro and geothermal sources are largely tapped.

A variety of renewable power sources are being studied, but wind is beginning to flourish, partially due to assistance from the Interministerial Committee for Prices, which in 1992 introduced a law providing premium prices for electricity from renewable sources.

Enel, the Italian utility, will soon have 800 MW of wind capacity with plans to add another 350 MW in the next couple of years. Edison ES has another 50 MW of wind capacity in Italy.

Wind potential in Sicily is estimated at 2000 MW and Enel Green Power is planning about 400 MW of wind power for the island.

> *For sure the mountainous terrain in Sicily provides good wind conditions. But at the same time you have several problems with transportation and erection of turbines because of access to the site, as the few roads that exist are narrow. Environmental and landscape issues are also to be taken seriously into account and for this reason flat coastal lands are almost out of limits, as are areas near to archaeological sites. So scouting for sites is really not so easy.*
>
> STEFANO SAVIO,
> KEY WIND DEVELOPER, ENEL

Enel has three wind farms in operation and about 200 MW in the process of obtaining permits. It takes about one year to obtain all the necessary licenses to begin construction, but the utility already purchased the land and established agreements with local municipalities, including paying the municipalities a percentage of the farm's income.

So far, the Italian public's attitude toward the wind projects has been mostly positive, but developers wonder whether attitudes will change with experience. Public attitude and visual impact of the wind farms are two major worries for developers there.

Offshore wind for Italy is doubtful for a couple of reasons. First, the Mediterranean Sea around the country is generally too deep for turbine siting. Second, tourism is an important part of the economy, and businesses and cities dependent on tourism would fight any offshore plans.

Case Study: Arklow Bank Project Siting 200 Turbines Offshore

Ireland has some of the best offshore wind resources in Europe and the Arklow Bank project will be the first of its kind in the Irish Sea. The Arklow Bank's project has planning consent for a 200-turbine facility with a nominal 500 MW capacity, which makes it the largest offshore project in the world (Fig. 3–7).

The turbines will be sited along the Arklow Sandbank, which is a stable structure capable of supporting the development. Arklow was selected as the preferred location for the wind park because of the existence of the sandbank with shallow water and the possibility of good ground conditions providing suitable foundation conditions at reasonable cost. The good winds also made the Arklow area an attractive proposition.

FIG. 3–7 GE's 1.5 MW WIND TURBINES AT ARKLOW (PHOTO COURTESY OF GE)

Throughout the development phase of the project, area residents expressed overwhelming support for the project. Open days were held in both Arklow and Courtown where national and local representatives, in addition to members of the public, were shown photomontages of the development, informed about progress, and advised of plans for its implementation.

A monitoring mast has been reading wind speeds and directions for several years to build a data set sufficient to select the optimum turbine type, layout, and power output to be determined. The mast is a 30-ton structure sitting in 5 m of water on the northern tip of the bank and is anchored by a steel monopile driven 15 m into the seabed. The mast extends 40 m above sea level.

As part of the preparations for the start of construction in 2003, a geophysical survey was performed on the sandbank and the undersea cable route to shore. The overall objective of the geophysical survey was to obtain:

- accurate bathymetric information in and around all installation points
- knowledge about obstacles, stones, boulders, wrecks, and/or archaeological remains at the sea floor
- knowledge about buried wrecks
- shallow geological information—especially to allow extrapolation of the results of geo-technical borings
- information about the possibility/ease of dredging for the cables

A geotechnical investigation, covering both the turbine locations and the cable route to shore, was conducted as well. The objective was:

- to accurately determine each soil interface
- to sample any soil layer
- to properly prepare the borehole with a minimum of disturbance for sampling and/or in situ strength testing to a total depth of 35 m below the seabed

The collected data sets were used to fine-tune the wind turbine foundation design for its summer 2003 delivery. An almost endless list of permits and agreements were needed before construction could begin, including a connection agreement with the local 38 kV distribution system and a switchgear substation approval.

Foundations are monopole design. A monopole is a tubular steel construction driven into the seabed to a depth of more than 30 m. Special pile driving hammers are used for this operation, which repeatedly impacts the poles until embedded to the required depth.

The project will feature seven of GE Wind Energy's 3.6 MW wind turbines, one of the largest wind turbine models developed at equipment selection time. The GE turbines represent the most advanced wind turbine technology available, with an increased generator size, a rotor diameter of 104 m and a hub height of 72 m. The first commercial prototype was unveiled by GE in 2002 and is currently installed in Barrax, Spain.

Case Study: Brazil's Hydro Crisis Unleashes Wind Demand

Brazil's energy industry is unusual in that it is a large, sprawling country and is powered overwhelmingly through hydroelectric resources. A long-term drought a few years ago, however, punctuated the country's need for a more diversified generation base. The country is now working to add more natural gas–fired generation, but the energy shortage has also spurred interest in wind energy development. A number of projects, totaling more than 1000 MW of wind energy, are being developed in Brazil (Fig. 3–8).

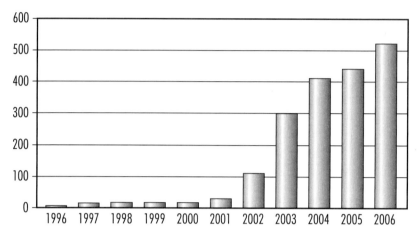

Fig. 3–8 Brazilian Wind Power (source: PEI, Frost & Sullivan)

Demand for energy is growing at 6% to 7% annually, so Brazil needs more capacity, and it doesn't want to use its domestic coal resources, which are high in ash and sulfur content. BP Solar has three solar projects underway in Brazil as well. Électricité de France is working to build U.S. $15 million in wind generation in Rio de Janeiro state. There is interest in adding offshore wind facilities as well.

4

Bioenergy

Biomass energy is derived from the energy stored in plants and organic matter. It is used to meet a variety of energy needs including generating electricity, heating homes, fueling vehicles, and providing process heat for industrial facilities.

Biomass also can be converted into transportation fuels such as ethanol, methanol, biodiesel, and additives for reformulated gasoline. Biofuels are used in pure form or blended with gasoline. Typical biomass fuels include the following:

Ethanol. Ethanol, the most widely used biofuel, is made by fermenting biomass in a process similar to brewing beer. Currently, most of the 1.5 billion gallons of ethanol used in the United States each year is made from corn and blended with gasoline to improve vehicle performance and reduce air pollution.

Methanol. Biomass-derived methanol is produced through gasification. The biomass is converted into a synthesis gas (syngas) that is processed into methanol. Of the 1.2 billion gallons of methanol annually produced in the United States, most is made from natural gas and used as solvent, antifreeze, or to synthesize other chemicals. About 38% is used for transportation as a blend or in reformulated gasoline.

Biodiesel. Biodiesel fuel, made from oils and fats found in microalgae and other plants, can be substituted for, or blended with diesel fuel.

Reformulated gasoline components. Biomass can also be used to produce reformulated gasoline components such as methyl tertiary butyl ether (MTBE) or ethyl tertiary butyl ether (ETBE).

Biomass resources include wood and wood wastes, agricultural crops and their waste byproducts, MSW, animal wastes, waste from food processing, and aquatic plants and algae. The majority of biomass energy is produced from wood and wood wastes, followed by MSW, agricultural waste, and landfill gases (*see* Table 4–1).

TABLE 4–1
BIOMASS USE BY TYPE

Fuel	Trillion Btus	Fuel	Trillion Btus	Fuel	Trillion Btus
Wood Energy		*Waste Energy*		*Other Biomass*	
Residential	407	MSW/Landfill Gas	434	Commercial	6
Commercial	43	Commercial	40	Industrial	90
Industrial	1,580	Industrial	104	Electric Power	21
Electric Power	140	Electric Power	290		
Wood Total	2,170	Waste Energy Total	551	*Alcohol Fuels*	
				Transportation	133

Dedicated energy crops. Fast growing grasses and trees grown specifically for energy production are also expected to make a significant contribution in the next few years.

Landfill gas plants collect methane gas (the primary component of natural gas) to run generators. Methane is a flammable gas produced from landfill wastes through anaerobic digestion, gasification, or natural decay. More than 100 power plants in 31 states burn landfill-generated methane.

Case Study: Landfill Gas-to-Energy Powers Iowa

At Metro Park East Sanitary Landfill near Des Moines, Iowa, a 6.4 MW landfill gas-to-energy project is one of many such projects in a nationwide Waste Management Corp. initiative. Waste Management has more than 60 Caterpillar generator sets (gensets) running on landfill gas and providing more than 48 MW of electricity to utilities across the country.

The Metro Park site was installed to beta test new technologies for use at other installations. With some of the most advanced technology, the facility has logged more than 99% up-time, including scheduled maintenance on eight Caterpillar G3516 SITA-LE gensets (Fig. 4–1).

Each genset is capable of producing 800 kW of continuous power. Since landfill gas can fluctuate from the normal 55% methane content, electronic control modules monitor the amount and quality of the intake fuel, optimizing the engines' performance.

FIG. 4–1 WASTE MANAGEMENT HAS INSTALLED GENSETS AT LANDFILLS ACROSS THE COUNTRY

Using advanced methane gathering methods, the landfill produces more than 3.2 million cubic feet (MMcf) of gas per day. An underground piping system spans 5.5 miles, connecting 70 eight-inch diameter well bores drilled to a depth of 86 ft. The electricity from this site is sold to the local utility for grid distribution.

The facility burns the equivalent of 112,000 barrels of oil annually in landfill gas. The methane gas, which contributes to ozone depletion when not properly disposed of, is effectively burned off in the on-site power plant.

Wood-related industries and homeowners consume the most biomass energy. The lumber, pulp, and paper industries burn their own wood wastes in large furnaces to heat boilers to supply energy needed to run factories. Homeowners burn wood in stoves and fireplaces to cook meals and warm their residences. Wood is the primary heating fuel in 3 million homes and is used to some degree in 20 million homes (Fig. 4–2).

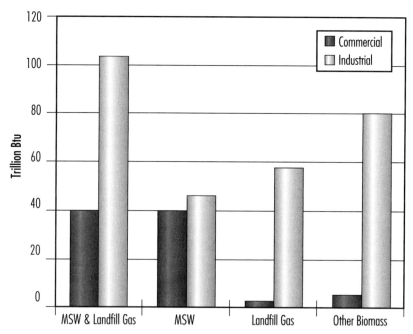

FIG. 4–2 WASTE ENERGY USE BY TYPE (SOURCE: DOE)

Biomass energy is used to make electricity, liquid fuels, gaseous fuels, and a variety of useful chemicals. Because the energy in biomass is less concentrated than the energy in fossil fuels, new technologies are required to make this energy resource competitive with coal, oil, and natural gas. Industry and agriculture need superior energy crops and cost-effective conversion technologies to expand the use of renewable biomass.

There are several ways to generate electricity from biomass, including the following:

Direct combustion. Biomass is burned to produce steam, the steam turns a turbine and the turbine drives a generator, producing electricity. Because of potential ash buildup and ensuing maintenance expense, only certain types of biomass materials are used for direct combustion.

Gasification. Gasifiers are used to convert biomass into a combustible gas, called biogas. The biogas is then used to drive a high-efficiency, combined-cycle gas turbine.

Pyrolysis. Heat is used to chemically convert biomass into pyrolysis oil. The oil, which is easier to store and transport than solid biomass material, is then burned like petroleum to generate electricity.

Biomass energy generates far fewer air emissions than fossil fuels, reduces the amount of waste sent to landfills, and decreases reliance on foreign oil. To increase use of biomass energy, dedicated energy crops must be developed, system efficiencies must improve, infrastructure to efficiently transport biofuels must be developed, and the cost of biomass energy must become more cost-competitive with fossil fuels.

Biomass potential

Biomass and bioenergy, such as agricultural products, forest and paper products, and waste-derived fuels satisfy about 4% of U.S. energy needs. This total includes more than 500 biomass-fired power plants with a combined capacity exceeding 7000 MW and many other cogeneration facilities.

The use of bioenergy products in the United States has been increasing about 2% annually since 1990. Freedonia Group projects demand for biomass energy and raw materials to reach $2.8 billion by 2005. Power generation is (and will remain) the primary end-use for biomass. Demand is expected to climb from 24 billion kWh in 2000 to more than 40 billion kWh in 2010.

Bioenergy is derived from waste materials that otherwise create unsightly messes. Converting waste to bioenergy also often prevents atmospheric pollution created when matter rots over time. Pictured in Figure 4–3 is a heap of newspaper, a prime source of bioenergy.

FIG. 4–3 BIOENERGY IS DERIVED FROM WASTE MATERIALS (COURTESY OF NREL)

Biomass-based energy production can provide various economic, environmental, and technical benefits. Converting these benefits into commercial reality, however, is not necessarily easy. There are at least four barriers to wider commercialization in the United States:

- low prices for natural gas, coal, and power
- little economic benefit for commercializing more efficient green technologies
- lack of industry familiarity with the technology options
- restrictions on the production tax credit so only closed-loop biomass projects using dedicated, plantation crops are eligible

Environmental and efficiency pressures along with the search for low-cost fuels will still likely support increased use of biomass for power production and thermal energy applications going forward.

In the near term, biomass-fired or co-fired power plants or cogeneration facilities will be fueled predominantly by waste biomass resources. Long-term, dedicated farm-grown biomass crops hold significant promise as a potentially large resource that gives biomass a way to reach 10% or more of U.S. electricity supply, up from the 1.5% of today. (If you include wood wastes and other readily collected residues, biomass jumps to 3% to 5%.)

> *With super high-yield crops, biomass could have a 10%, even 20 to 25% role in future electricity supply. Such closed-loop, farm-grown biomass is both the hope and the challenge for biomass energy.*
>
> EVEN HUGHES, EPRI

The challenge is the search for higher yields and lower costs, especially harvesting costs. While dedicated biomass crops offer a promising pathway toward carbon mitigation, the associated technical and economic constraints mean the first steps on that pathway will likely focus on wastes and residues as the biomass fuel source. Fossil-derived expertise and fossil power plant assets provide an excellent starting point from which to launch a new round of growth in biomass power generation.

The growing worldwide interest in gasification due to its fuel flexibility, environmental performance, and high efficiency has logically led to interest in biomass gasification. Because of its unique properties, including high reactivity, low ash, low sulfur, and high volatile matter content, biomass is very different from fossil fuels and requires a different approach to optimize performance.

Building on efforts begun in the 1980s by Battelle and the U.S. DOE, FERCO is pursuing commercialization of a unique biomass gasification process called Silvagas. FERCO has been testing a commercial-scale demonstration unit of 200 tons-per-day at the 50 MW wood-fired McNeil Generation Station in Burlington, Vermont (Fig. 4–4).

Fig. 4–4 The McNeil Station in Vermont
(photo courtesy of FERCO)

The SilvaGas process converts biomass into a medium-Btu gas of 450 to 500 Btu/cf that can directly displace natural gas as fuel in power production, steam production, or heating. Two circulating fluidized bed reactors function as the primary process vessels, one as the gasifier and one as the combustor. Sand at 1800°F circulates between the two vessels, thermally breaking down the highly reactive biomass and conveying char from the gasification reactor to the process combustor.

The McNeil demonstration has completed several extended test campaigns since its installation in 2000. Gas quality has been consistent and reproducible from a range of fuels.

While gas cleanup is complicated, the absence of sulfur in biomass feedstocks eliminates one of the main issues faced by coal gasifiers. Removing particulate matters effectively is important, and particulate levels of less than 5 parts per million (ppm) can be achieved using a conventional, off-the-shelf wet scrubbing process. Condensable hydrocarbon removal is more difficult but can be achieved through a cracking operation similar to those used in the petroleum industry.

Extremely low contaminant levels are particularly important for applications where biogas will be used to fire gas turbines because the turbines are more easily damaged by solid contaminants.

Production costs for the SilvaGas are competitive with natural gas at the burner tip because natural gas transportation costs are avoided.

Case Study: PGE Makes Green Power with Cows

Portland General Electric (PGE) has enlisted Oregon's dairy industry to tap a renewable source of green power from cows. The process produces power from renewable resources while capturing and using methane gas that would otherwise be emitted to the atmosphere.

PGE is running two pilot projects—one of 100 kW and one 4 MW. The 500 cows at the Cal-Gon Farms in Salem produce manure converted to methane in a digester PGE installed at the site. The methane fires a Wakesha VSG 11 GSID 6-cylinder engine to produce 100 kW for the utility's distribution system.

> *Essentially, a digester is like a big mechanical stomach. It continues the same process that takes place in the cow's stomach, breaking down organic material. We're just capturing the byproduct that process produces, methane.*
>
> JEFF COLE,
> BIOGAS PROGRAM MANAGER, PGE

In addition to cutting methane emissions, the process reduces odor from the dairy property and produces useful byproducts such as nitrogen-rich liquid fertilizer and sanitary fiber that can be used to make various materials for the nursery industry. PGE sells the majority of the digester fiber to a local company. The digester process should also help dairy farmers comply with stricter federal regulations limiting phosphate levels in the soils.

Anaerobic digestion got its start during the energy crisis of the 1970s, but most digesters failed because farmers weren't trained to maintain the complex machines, or simply lost interest once energy prices decreased. There are only about 30 digesters currently operating on commercial livestock farms nationwide and only about 15 on dairy farms, according to the EPA.

But recent energy shortages, especially in the west, renewed interest in the technology. Oregon's dairy industry is actually the smallest in the northwestern states. California's dairy industry is the largest in the West. California has provided $15 million for development of anaerobic digestion technology. A $10 million grant program approved recently also helps dairies pay for digesters.

The Cal-Gon Farms digester is unique because it is entirely paid for and maintained by PGE, which supplies 730,000 Oregon homes and businesses with power. The experimental program came from a partnership between the Oregon Dairy Farmers Association and PGE. It marks the first time a utility has coordinated with farmers to develop an economically feasible digester at the utility's expense.

PGE plans to use the lessons learned from the Cal-Gon Farms project to develop additional agricultural partnerships. A far larger project with Three-mile Canyon Farm is in the works, expected to enter service in 2003. It will be the largest agricultural digester gas installation in the nation. About 20,000 cows are available at that facility, which will generate 4 MW.

Biomass for heat and power

District Energy St. Paul (DESP) provides heating and cooling services to more than 150 buildings and about 300 single-family residences in downtown St. Paul, Minnesota. In the early 1990s, several factors convinced the nonprofit company who owns and operates the district heating and cooling plant to develop a new biomass combined heat and power plant.

First, the state required the local utility to obtain a certain percentage of its electricity from renewable resources. Second, DESP wanted to reduce the environmental impact from its use of coal. A biomass plant could cut coal use by up to 80% and eliminate 280,000 tons of CO_2 emissions per year. Third, St. Paul and its surroundings had plenty of local biomass supplies to feed the plant. The area produces more than 600,000 tons of wood waste each year.

The $55 million facility entered service at the close of 2002 with a 310,000 lb/hr vibrating grate boiler from Foster Wheeler. The vibrating grate technology was chosen rather than fluidized bed technology for better overall economics, including lower operating and maintenance costs and greater flexibility in controlling combustion air distribution.

The boiler feeds steam at 1280 psi and 950°F to a 37.4 MW steam turbine-generator, with a net output of 33 MW. The plant produces hot water at 190° to 250°F depending on outside temperatures for distribution to customers, replacing about 80% of the thermal energy output provided by previous equipment, and producing about 25 MW of power in the winter to be sold to the local utility. In the summer the plant can produce up to 33 MW of power due to lower heating demands. The facility will receive a price higher than market price for the electricity it provides under the state renewable energy mandate.

Back for the future

Many of the older biomass power plants were shut down in recent years as their power purchase contracts either expired or were bought out. With the rising demand for electricity, however, some are being restarted. Energy Products of Idaho (EPI) refurbished and recommissioned one such facility, the Madera power plant near Fresno, California, to provide power to the state grid.

The original facility, commissioned in 1988, was mothballed in early 1995 when PG&E bought its power purchase agreement. In response to the California energy crisis in 2000, Madera Power, a subsidiary of EPI, bought the plant and started the retrofit project.

The plant uses EPI's patented atmospheric fluidized bed boiler, where biomass feedstocks from the San Joaquin Valley and as far away as San Francisco and Los Angeles are combusted. Heat is recovered in a forced-circulation boiler, and steam is fed to a steam turbine, which drives a 28.5 MW generator.

The Madera plant now operates 24 hours a day, seven days a week, except for maintenance outages or forced shutdowns. Total cost of the purchase and renovation was $650/kW, which is about one-third the cost of a new facility. Madera management invested in several optional upgrades to boost reliability and economic performance of the facility. The plant heat rate has averaged about 18,000 Btu/kWh since coming back on-line.

Biogas potential

Biogas is showing potential as a retrofit component to utility coal-fired power plants. Biomass gasification technology can use any material of biological origin, including wood waste, agricultural waste, or yard waste as fuel. The energy in the solid biomass waste is converted into a low Btu biogas. A study conducted by Black & Veatch and EPI shows that using biomass gasification at existing power plants can provide green power at a much lower cost, when compared to the expense of a new biomass plant.

Substituting biogas for part of the coal at a coal-fired plant can significantly reduce nitrogen oxide (NO_x), sulfur dioxide (SO_2), mercury (Hg), CO_2, and other pollutants, especially when biogas is used in a reburn configuration. Another potential advantage is the low cost of biomass fuels, which improves profitability.

> *This technology represents one of the lowest-cost renewable energy sources we have evaluated to date. That alone should make it an option for power providers to consider when they develop their strategic renewable energy plans.*
>
> RYAN PLETKA, PROJECT MANAGER, EPI

Availability of good biomass fuel nearby is important when determining feasibility of such a system. Project feasibility can also be affected by new and existing emissions regulations and status of renewable energy incentives available for a site from federal and state governments.

The study speculates that incentives for such projects could provide improved environmentally safe operations that would earn additional revenue for the facility owner.

> *The ideal project location would be a utility plant with a combination of high coal costs, access to low cost biomass or waste material, looming NO_x reduction requirements, and state mandates or incentives for renewable energy. There are several states in the eastern half of the country that have all these features. However, the presence of any of these is potentially enough to ensure project viability. Additional credits from reduction of NO_x, SO_2, and other emissions only serve to improve the economics of the project.*
>
> PLETKA

The study found that a *green premium* of $.014 to $.018 cents/kWh is enough to guarantee economic viability at a proposed site. Such a subsidy could come through retail green power programs, sale of renewable energy credits on the wholesale market, or expanded tax incentives for renewable generation. Even without incentives, biomass fuel expense of $.80 cents/thousand British thermal units (Mbtu) less than coal fuel expense is enough savings to justify a project.

In 1973 EPI provided the first fluidized bed combustion (FBC) system capable of using biogas from waste biomass in the United States. The facility, in Coeur D'Alene, Idaho, was installed for Idaho Forrest Industries.

EPI's biomass gasifier add-on for coal-fired boilers is a unique system that combines the production of green, renewable energy with effective NO_x reduction in a single unit. The normal problems and disadvantages inherent with directly co-firing biomass in coal boilers, such as excessive wear on pulverizers, fouling and slagging of tubes, ash contamination, etc., are completely eliminated or minimized through the use of EPI's gasifier.

PAT TRAVIS,
BUSINESS DEVELOPMENT MANAGER, EPI

Biopower basics

The energy stored in biomass (organic matter) is called bioenergy. Bioenergy can be used to provide heat, make fuels, and generate electricity. Wood, which people have used to cook and keep warm for thousands of years, continues to be the largest biomass resource. Today there are also many other types of biomass we can use to produce energy. These biomass resources include residues from the agriculture and forest industries, landfill gas, aquatic plants, and wastes produced by cities and factories (Fig. 4–5).

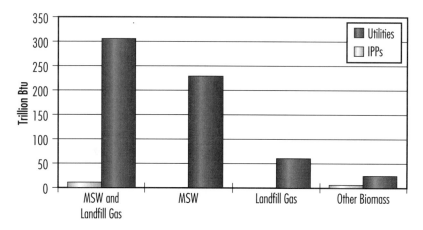

FIG. 4–5 WASTE ENERGY USE IN ELECTRIC POWER GENERATION (SOURCE: DOE)

Because they come from organic matter, biomass resources are renewable. For example, many biomass resources are replenished through the cultivation of fast-growing trees and grasses. As these trees and grasses grow, they remove CO_2—a major greenhouse gas—from the atmosphere. This is important because bioenergy, like fossil fuels, can produce CO_2. However, the net emission of CO_2 from bioenergy will be zero as long as plants continue to be replenished.

Hundreds of U.S. power plants use biomass resources to generate about 65 billion kWh of electricity each year. The wood and paper products industries generate and use about two-thirds of this power. Solid wastes from cities fuel most of the remaining biopower plants, providing enough electricity to meet the needs of nearly seven million Americans.

Biopower plants come in all sizes. Today's biopower plants have a combined capacity of about 10.3 GW, which is about 1.4% of the nation's total electrical generating capacity. However, with better technology and expanded use of biomass resources, the nation could generate as much as four-and-a-half times more biopower by 2020.

Of all the forms of renewable energy, only hydropower produces more electricity than bioenergy does. Like hydropower, biopower is available 24 hours a day, seven days a week. Other forms of renewable energy, such as solar or wind power, have lower availability since they are produced only when the sun shines or the wind blows.

Several types of biopower systems are currently in use or under development. These systems include direct combustion, co-firing, gasification, and small modular systems.

Direct combustion. Direct combustion involves the burning of biomass in a boiler to produce steam. The pressure of the steam then turns a turbine attached to an electrical generator, which makes electricity. Coal-fired power plants employ similar technology but use fossil fuel in their boilers. Most of today's biopower plants use a direct combustion system. Researchers are evaluating other advanced processes that are even more efficient than direct combustion.

Co-firing. Co-firing systems can burn up to 15% biomass when mixed with coal in some boilers. Co-firing biomass with coal reduces emissions. Many existing coal plants could use a co-firing system with only a few modifications. Therefore, this system has a significant potential for growth in the near future. To make co-firing biomass more attractive to power companies, researchers are investigating improvements to the co-firing process and better technologies for minimizing emissions.

Gasification. Engineers are developing new technologies to produce biogas from biomass. Biogas consists of methane (found in natural gas) together with hydrogen, and other gases. Some new gasification technologies make biogas by heating wood chips or other biomass in an oxygen-starved environment. A second way to make biogas is to let landfills do the work. As paper and other biomass decay inside a landfill, they naturally produce methane. Methane can be recovered from landfills by drilling wells into the landfill and piping the gas to a central processing facility for filtering and cleaning.

Pyrolysis. Researchers are also investigating a smoky-colored, sticky liquid that forms when biomass is heated in the absence of oxygen, called pyrolysis oil. This liquid can be burned like petroleum to generate electricity. Petroleum, however, is almost never used any more to generate electricity. There's a greater need to use petroleum as a source of gasoline, heating oil, and petrochemicals. Because pyrolysis oil can also be refined in ways similar to crude oil, it may be more valuable as a source of biofuels and biobased products than for biopower generation. Unlike direct combustion, co-firing, and gasification, this technology is not yet in the marketplace.

Modular systems. Researchers are particularly interested in improving small systems sized at 5 MW or less. These so-called modular biopower systems can use direct combustion, co-firing, or gasification for power generation. They are well suited for generating biopower from locally grown resources for small towns, rural industries, farms, and ranches.

Transportation applications

Biomass is the only renewable source of transportation fuels. These renewable fuels, called biofuels, produce fewer emissions than petroleum fuels. Biofuels also can help us reduce U.S. dependence on foreign sources of fossil fuels. Biofuels can also stimulate growth in industry and in rural areas, making farming and forestry more profitable. The two main biofuels are ethanol and biodiesel.

Ethanol

Ethanol is an alcohol, the same found in beer and wine. It is made by fermenting any biomass high in carbohydrates (starches, sugars, or celluloses) through a process similar to brewing beer. Ethanol is mostly used as a fuel additive to cut down a vehicle's carbon monoxide (CO) and other smog-causing emissions. Flexible-fuel vehicles, which run on mixtures of gasoline and up to 85% ethanol, are available.

Industry currently makes ethanol from the starch in grains—such as wheat, corn, or corn byproducts—in a process similar to brewing beer. Each year, more than 1.5 billion gallons of ethanol is blended with gasoline to improve vehicle performance and reduce air pollution.

Most gasoline blends contain about 10% ethanol and 90% gasoline. This mixture works well in cars and trucks, those you see on the road everyday, designed to run on gasoline. In addition, fuel containing 85%

ethanol is available, primarily in the Midwest. This fuel, called E85, can be used in flexible fuel vehicles. Flexible fuel vehicles can run on E85, straight gasoline, or any mixture of the two. Each year, automobile manufacturers produce more than 700,000 flexible fuel vehicles.

Researchers are investigating technologies for making ethanol from the cellulose (fiber) component in biomass, like MSW and agricultural residues left in the field after harvest. This type of ethanol is called bioethanol. Bioethanol reduces exhaust emissions from CO and hydrocarbons. In addition, by displacing gasoline components like sulfur, bioethanol helps reduce the emissions of toxic effluents from automobiles.

> **Ethanol fun fact**
>
> Ethanol's history as a transportation fuel goes back many years to Henry Ford and other transportation pioneers. In the 1880s, Ford built one of his first automobiles—the quadricycle—and fueled it on ethanol. Early Ford Model Ts had a carburetor adjustment that could allow the vehicle to run on this fuel, which would be produced by America's farmers. Ford's vision was reportedly to build a vehicle that was affordable to the working family and powered by a fuel that would boost the rural farm economy.

Biodiesel

Biodiesel is made by combining alcohol (usually methanol) with vegetable oil, animal fat, or recycled cooking greases. It can be used as an additive to reduce vehicle emissions (typically 20%) or in its pure form as a renewable alternative fuel for diesel engines. Industry produces about 20 million gallons of biodiesel from recycled cooking oils and soybean oil.

Like ethanol, biodiesel is primarily used as a fuel blend. Diesel blends usually consist of 20% biodiesel with 80% petroleum diesel. This mixture runs well in a diesel engine and does not require engine modifications.

Biodiesel is not yet widely available to the general public. Some federal, state, and transit fleets, as well as tourist boats and launches, use blended biodiesel or pure biodiesel. Industry is looking at using biodiesel in circumstances where people are exposed to diesel exhaust; in aircraft to control pollution near airports, and in locomotives with unacceptably high emissions. Biodiesel may increase NO_x emissions but it reduces CO, particulates, soot, hydrocarbons, and toxic emissions when compared to pure, petroleum diesel.

Biodiesel market

The use of biodiesel has grown dramatically during the last few years. The Energy Policy Act (EPAct) was amended in 1998 to include biodiesel fuel use as a way for federal, state, and public utility fleets to meet requirements for using alternative fuels. That amendment started the sharp increase in biodiesel users, which includes the U.S. Postal Service and the U.S. Departments of Energy and Agriculture.

Countless school districts, transit authorities, national parks, public utility companies, and garbage and recycling companies also use the fuel. The biodiesel industry has achieved this level of growing success largely on its own, without the help of government subsidies.

According to the American Biofuels Association, with government incentives comparable to those provided for ethanol, biodiesel sales could reach about 2 billion gallons per year, or about 8% of highway diesel consumption. At this level of market penetration, biodiesel would probably be used in bus fleets and heavy-duty trucks (primarily in blends with fossil diesel at the 20% level), marine vessels such as ferries, construction and agricultural vehicles, home heating oil systems, and electric generation.

Biodiesel currently costs between $1 and $2 per gallon and could compete with low-sulfur diesel fuels. Feedstock costs account for a large percent of direct production costs, including capital cost and return. It takes about 7.3 pounds of soybean oil, which costs about $.20 cents per pound, to produce a gallon of biodiesel. Feedstock costs alone are at least $1.50 per gallon of soy biodiesel.

Fats and greases cost less and produce less expensive biodiesel, sometimes as low as $1.00 per gallon. The quality of the fuel is similar to soy biodiesel fuel. Sophisticated feedstock blending strategies will begin to address consumer requests for low NO_x fuels for summer and good cold flow fuels in the winter.

Under the mustard seed program, oil can be produced today for approximately $.10 cents per pound, and the total cost of producing mustard biodiesel is around $1.00 per gallon. The goal with mustard seeds is to develop a growing demand for organic pesticides made from mustard meal, which will provide the primary incentive to farmers and crushers. The mustard oil is a low value waste product because it's inedible.

Case Study: Savannah River Switches to Biodiesel

Savannah River switched all its diesel engines to biodiesel in January 2001. The DOE site has about 190 diesel vehicles and more than 430 pieces of portable equipment that are powered by diesel engines—all burn biodiesel. No equipment conversions were needed for the fuel switch.

The Savannah River Site uses about 35,000 gallons of B20 (20% biodiesel/80% diesel) monthly, earning Alternative Fuel Vehicle credits required by the EPAct.

> *We haven't had a single complaint or problem. The biodiesel is similar in terms of fuel consumption and horsepower and it integrates easily into our existing fueling tanks. B20 costs us only about 20% more than standard diesel fuel, and we feel the cost is justified. It provides greater lubricity, which we believe will eventually lead to decreased maintenance costs, and it reduces most regulated emissions significantly.*
>
> CHRIS GOODMAN,
> OPERATIONS OFFICER, SAVANNAH RIVER

DOE is working to expand its use of biodiesel.

> *The Department of Energy will rely on biodiesel blends to achieve a large portion of the 20% reduction in petroleum use called for by 2005 by Executive Order 13149.*
>
> LEE SLEZAK, DOE

Case Study: Florida, Alabama Utilities Laud Biodiesel

For Florida Power & Light (FPL) and Alabama Power, biodiesel has been the alternative fuel of choice. Both utilities started using biodiesel on a trial basis, but were so pleased with the fuel that they now use it as much as possible.

> *When we began with B20, we did it as a test and part of that test was to find out if our operators noticed the difference. They didn't. We had no complaints at all.*
>
> RICHARD HARPER,
> ALABAMA POWER

Finding the right supplier is important for a successful biodiesel program.

> *Make sure to find a supplier who can work with you to meet your needs. For example, we store no fuel at our facilities. Each vehicle is fueled by our supplier each evening and they have to be willing to bring the blended B20 from their facility.*
>
> TIM CALHOUN, FPL

Beyond earning EPAct credits, there are other benefits from biodiesel.

> *It definitely increases the lubricity of the fuel and our tests show it has reduced emission of hydrocarbons, particulate matter, and CO_2. We've also found it really does clean up the fuel system.*
>
> CALHOUN

Alabama Power uses the fuel in 30 to 40 trucks in the Birmingham area. FPL selected the fuel for several reasons, including availability.

> The fuel has to be versatile in that some of the vehicles are older—up to 15 years—and others were new when we started with biodiesel. We have a variety of engines and a variety of uses, and biodiesel has worked well in each case. Being a southern state, the infrastructure for propane or CNG alternatives just isn't there.
>
> CALHOUN

Other biobased products

Other biofuels include methanol and reformulated gasoline components. Methanol, commonly called wood alcohol, is currently produced from natural gas, but could also be produced from biomass. There are a number of ways to convert biomass to methanol, but the most likely approach is gasification. Gasification involves vaporizing the biomass at high temperatures, then removing impurities from the hot gas and passing it through a catalyst, which converts it into methanol.

Most reformulated gasoline components produced from biomass are pollution reducing fuel additives, such as MTBE and ETBE.

Researchers have discovered that almost any product that can be made with fossil fuels can be made from biomass. The difference between a chemical derived from plants and an identical chemical made from petroleum is simply the origin. Biobased products often require less energy to produce than petroleum-based products. In addition, they can be made from otherwise useless waste material.

The United States produces more than 300 billion pounds of biobased products each year, not counting food and feed. Biobased products include plastics, cleaning products, natural fibers, natural structural materials, and industrial chemicals made from biomass. Such chemicals are sometimes referred to as *green* chemicals because they are derived from a renewable resource.

Phenol is an example of a green chemical. Typically made from coal tar, phenol can also be extracted from pyrolysis oil. Wood glues, which are used to make plywood, molded plastic, and foam insulation, are some of the materials that can be made with phenol.

Scientists have discovered how to release the sugars making up starch and cellulose in plants not only to manufacture biofuels but also a number of biobased products, including:

- antifreeze
- brake fluid
- glues
- softeners
- artificial sweeteners
- acids used in making cheese and soft drinks
- biodegradable plastics
- food thickeners such as xanthan gum
- gels for toothpaste, medicines, and paints

Carbon monoxide and hydrogen are also important building blocks for biobased products. When biomass is heated with a small amount of oxygen present, these two gases are produced in abundance. Scientists call this mixture biosynthesis gas. Once it is cleaned up, biosynthesis gas can be used to make:

- electricity
- alcohol fuels
- antifreeze
- acids used in making photographic films, textiles, plastics, and synthetic fabrics
- sulfur-free gasoline
- sulfur-free diesel fuels
- other products made with fossil fuels, including plastics

Biobased products are so varied it's unlikely industries in the future will limit themselves to making just one of them. Rather, biorefineries could become commonplace (*see* Table 4–2).

TABLE 4–2
INDUSTRIAL BIOMASS USE

Industry	For Power	For Heat
Agriculture, Forestry, Mining	3	7
Manufacturing	376	1,333
Food and Food Products	3	46
Lumber	20	247
Paper and Paper Products	351	989
Chemicals	1	21
Other	0	30
Unspecified Business Type	0	103

Source: DOE

Biomass at work

Biomass resources are plentiful and varied throughout the country. They are primarily wastes, food crops, and energy crops. Wood and wood waste are the biggest sources of bioenergy in the United States. Wood-powered energy is created mostly by lumber and paper industries, which use their waste products to create both heat and power for use in industrial facilities (Fig. 4–6).

FIG. 4–6 WOOD AND WOOD WASTE (PHOTO COURTESY OF NREL)

In the Pacific Northwest and the Southeast, the forest products industry uses its wastes and residues to make electricity and heat for its own operations. Instead of filling up a landfill, sawdust, bark, paper pulp, wood shavings, scrap lumber, wood dust, and paper provide low-cost bioenergy. In Hawaii a plant is using *bagasse* (a fibrous residue from sugar cane processing) to make particleboard.

In the Midwest, farmers grow corn and soybeans for ethanol fuels and bioproducts. A South Dakota firm sells truck bed liners made from soybeans. A Minnesota firm makes shrink wrap, clothing, candy wrappers, cups, food containers, home and office furnishings, and other biodegradable products from a chemical building block derived from corn starch. A consortium of farmers, businesses, and utilities in Iowa is growing 4000 acres of switchgrass as an energy crop for co-firing with coal in utility boilers.

A similar consortium in the Northeast is growing hybrid willow trees as energy crops, also for co-firing with coal. A number of cities in the Northeast generate electricity from their biomass-rich solid wastes instead of burying them in landfills. A utility in Vermont is experimenting with a new system to make biogas from wood chips.

Every day, a fast-food corporation delivers hamburgers all over the country in clamshell containers made, in part, from starches recovered from the production of French fries and potato chips.

The use of these resources is laying the foundation for future bioenergy use. However, if we want to increase our bioenergy resources and lower the costs of producing them, we must rely more on energy crops and less on food crops. As our understanding of agricultural science grows, we'll be able to grow more and better energy crops. Potential energy crops include poplars, willows, switchgrass, alfalfa stems, and sweet sorghum.

Compared to conventional farming, energy crops require less fertilizer and fewer chemicals to control weeds and insect pests. With sustainable farming practices, energy crops could be used to prevent erosion and to protect water supplies and quality. Researchers are developing perennial grass and tree crops with life expectancies of 7 to 10 years after planting. Research has shown that soil carbon, one indicator of soil quality, increases measurably under energy crops in as few as 3 to 5 years. These crops can potentially restore the cultivation and water-holding capacity of soil degraded by intensive crop production. In all these ways, energy crop farming can help preserve cropland for future generations.

World Views

Worldwide, biomass is the fourth largest energy resource after coal, oil, and natural gas. It is used for heating (such as wood stoves in homes and for process heat in bioprocessing industries), cooking (especially in many parts of the developing world), transportation (fuels such as ethanol) and increasingly, for electric power production. There are estimates of about 35,000 MW of installed capacity using biomass worldwide, with about 7000 of that in the United States. Most of this capacity is in the pulp and paper industry in combined heat and power systems.

Recent studies indicate that additional (presently unused) quantities of economically available biomass may exceed 39 million tons per year in the United States—enough to supply about 7500 MW of new biopower, or a doubling of the existing biopower capacity. This could supply the amount of electricity used each year by the residential customers in all six New England states combined. Economic availability of energy crops and crop residues could increase this quantity tenfold.

Worldwide biopower generation is expected to grow to more than 30,000 MW by 2020. In many countries, local environmental conditions and global climate change concerns are further stimulating the demand for clean energy. Off-grid, modular systems offer the most viable international market opportunity for biopower. Developing countries are the top markets because they meet several criteria:

- rapid economic growth
- burgeoning demand for electricity
- mounting environmental problems
- need for rural electrification
- need for reliable electricity
- significant agricultural/forestry residues

China and India are considered to be the prime candidates. Estimates show that by 2015, China will have between 3500 and 4100 MW of biopower capacity and India will have between 1400 and 1700 MW. This is a sharp rise from their current levels of 154 MW and 79 MW, respectively. These two countries may also be good targets for co-firing operations because they have many older coal-fired power plants where biomass co-firing could be used to economically improve environmental performance. Other countries that show promising growth for a variety of biopower systems are Brazil, Malaysia, Philippines, Indonesia, Australia, Canada, England, Germany, and France.

Case Study: Biomass Use in Bangladesh

On many levels, Bangladesh is a country that is ideally suited for the development of small-scale biomass energy systems. Because the economy is largely dependent on agriculture, the residues needed for such projects are available. Approximately 75% of the 130 million people in the country live in rural areas, and for all practical purposes they are not able to benefit from the national electricity transmission grid.

The country is relatively poor, with a per capita annual income of less than U.S. $300 (Neighboring India, has an annual per capital average of about U.S. $500). As a result, it is difficult to attract the investment needed to expand the national energy infrastructure.

The lack of infrastructure in Bangladesh's rural areas has resulted in an increase in the migration of rural populations to the country's urban areas, putting enormous pressure on urban infrastructures ill equipped to deal with the additional population. As a result, the Bangladeshi government is interested in finding economical ways to bring electricity to the rural areas, both to improve economic development and to stem the migratory trend. Small-scale renewable energy systems fueled by biomass may offer Bangladesh a way to accomplish these goals.

The technologies most popular in terms of development are biogas digesters running on animal or human wastes, turning agricultural wastes into solid fuel briquettes (similar to charcoal), and direct combustion of agricultural waste for household cooking. The main need for energy in rural Bangladesh is for cooking, although biomass is also used as housing material and animal feed. A limited amount of biomass is used as feedstock for recycled paper and in pulp mills.

The main sources of biomass in Bangladesh include rice husks, jute stalks, sugarcane stalks, and peanut shells. The patterns of biomass usage in developing countries such as Bangladesh could not be more different from those in industrialized countries like the United States. In industrialized countries there is an abundance of waste biomass material that has only been used once and is contained in landfills, forests, or agricultural lands. A waste stream that may be attractive in the United States, like MSW, is fraught with problems in developing countries. In the United States, wastes are carefully entombed in landfills and generally left undisturbed. In developing countries, entire communities of rag pickers live on and alongside the dumps and earn their living by scavenging materials and selling them to small industries turning them into a myriad of products ranging from combs to shoes to paper. Consequently, attempts to divert streams of MSW in developing countries can affect entire classes of people and the small industries depending on them.

Although large quantities of *waste* are generated in a country like Bangladesh due to the agricultural nature of the economy, relatively little of that biomass may be available for use in energy generation. As long as competing uses of biomass material fetch a higher price, or are easier to accomplish, the material will find use in nonenergy applications. Two examples illustrate the opportunities and pitfalls for biomass commercialization in a developing country like Bangladesh.

A thriving business in Bangladesh is biomass briquetting or *densification*. Briquetting processes require heat and pressure to produce fuel pellets from rice husks and wood chips. There are approximately 900 briquetting machines in operation in Bangladesh, the overwhelming majority of which are locally manufactured. Briquettes have become popular as a fuel for heating urban hotels and tea shops. In addition, briquettes are in demand as a fuel for melting bitumen, which is used in road paving operations. Brick manufacturing industries also can use the briquettes as a fuel in their ovens. Overall, the prospects for growth of this industry in Bangladesh look bright.

Another example of biomass use in Bangladesh is biodigesters. Unlike briquetting, biodigesters have a mixed record of success. In Faridpur District, a school with about 350 students and 50 staff members uses a biodigester to generate a methane-based cooking fuel. Sludge from the digester is used for fertilizer. The replacement cost for a plant of this type is estimated to range from $515 to $825. Bangladesh's local government engineering department provided the initial funding and paid for the entire system. The school pays for the operating and maintenance costs, which have been negligible. Although there have been successful installations of other biodigesters in the community, the school has not expanded its own biodigester program. The principal barriers to further commercialization of the technology are high capital costs and lack of financing options.

Despite economic and availability hurdles, it is expected that biomass will continue to play a key role in supplying the energy needs of Bangladesh. The major question is how quickly more efficient biomass-using technologies can be introduced to allow the people of Bangladesh to obtain maximum benefit from the resource.

Geothermal Energy

Miles beneath the earth's surface lies one of the world's largest energy resources—geothermal energy. Our ancestors used geothermal energy for cooking and bathing since prehistoric times. Today, this enormous energy reservoir supplies millions of people with clean, low-cost electricity.

Geothermal energy is the heat contained below the earth's crust. This heat, brought to the surface as steam or hot water, is created when water flows through heated, permeable rock. It's used directly for space heating in homes and buildings or converted to electricity (Fig. 5–1).

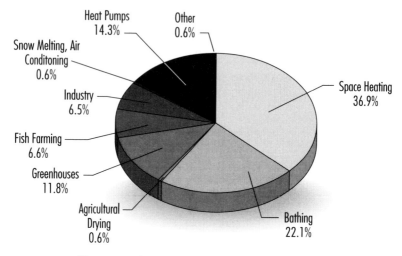

Fig. 5–1 Geothermal Heat, Direct Use
(source: Geothermal Energy Association [GEA])

Most of the country's geothermal resources are located in the western United States. The United States has more than 2700 MW of geothermal electric power capacity, coming from natural, occurring steam and hot water. This only scratches the surface of the potential electricity production of geothermal resources. Geothermal resources come in five forms:

- hydrothermal fluids
- hot dry rock
- geopressured brines
- magma
- ambient ground heat

Of these five, only hydrothermal fluids have been developed commercially for power generation. Three technologies can be used to convert hydrothermal fluids to electricity. The type of conversion used depends on whether the fluid is steam or water, and its temperature (Fig. 5–2).

FIG. 5–2 HOT SPRINGS STEAM

Steam

Conventional steam turbines are used with hydrothermal fluids that are wholly or primarily steam. The steam is routed directly to the turbine, which drives an electric generator, eliminating the need for the boilers and conventional fuels to heat the water.

High-temperature water

For hydrothermal fluids above 400°F that are primarily water, flash steam technology is usually used. In these systems, the fluid is sprayed into a tank held at a much lower pressure than the fluid, causing some of the fluid to rapidly vaporize, or flash, to steam. The steam is used to drive a turbine, which again, drives a generator. Some liquid remains in the tank after the fluid is flashed to steam, if it is still hot enough. This remaining liquid can be flashed again in a second tank to extract even more energy for power generation.

Moderate-temperature water

For water with temperatures less than 400°F, binary-cycle technology is generally most cost effective. In these systems, the hot geothermal fluid vaporizes a secondary, or working fluid, which then drives a turbine and generator.

Steam resources are the easiest to use, but are rare. The only steam field in the United States commercially developed is the Geysers, in northern California. The Geysers began producing electricity in 1960. It was the first source of geothermal power in the country and is still the largest single source of geothermal power in the world (Fig. 5–3).

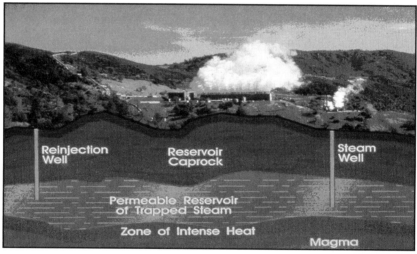

FIG. 5–3 THE GEYSERS (COURTESY OF CALPINE)

Hot water plants, using high- or moderate-temperature geothermal fluids, are a relatively recent development. However, hot water resources are more common than steam. Hot water plants are now the major source of geothermal power in both the United States and the world (Fig. 5–4).

FIG. 5–4 A MODERN GEOTHERMAL POWER PLANT

Today's hydrothermal power plants with modern emissions controls have minimal impact on the environment. The plants release little or no CO_2. Geothermal power plants are very reliable when compared to conventional power plants, e.g., new steam plants at the Geysers are operable more than 99% of the time.

In some parts of the world, geothermal systems are cost competitive with conventional energy sources. It is anticipated that as technology improves, the cost of generating geothermal energy will decrease.

Power production

More than 2700 MW of electric power capacity produced in the United States originates under the earth's surface as naturally occurring steam and hot water. This is enough to provide electricity to 3.5 million homes. It only scratches the surface of the potential electricity production of geothermal resources (Fig. 5–5).

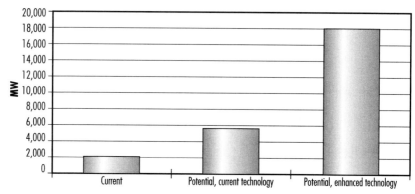

FIG. 5–5 U.S. GEOTHERMAL POTENTIAL (SOURCE: GEA)

Geothermal electricity is clean, reliable, and cost effective. Geothermal resources represent an abundant, secure source of energy. Geothermal resources are currently being used to generate electricity in a handful of U.S. states (*see* Table 5–1).

TABLE 5–1
STATES CURRENTLY USING GEOTHERMAL RESOURCES

Alabama	Georgia	Nevada	Utah
Alaska	Hawaii	New Mexico	Virginia
Arizona	Idaho	New York	Washington
Arkansas	Louisiana	Oregon	West Virginia
California	Mississippi	South Dakota	Wyoming
Colorado	Montana	Texas	

Source: Geo-Heat Center, Oregon Institute of Technology

Today's hydrothermal power plants with modern emissions controls have minimal impact on the environment. The plants release little or no CO_2. In fact, electricity produced from geothermal resources in the

United States displaces the emission of 22 million tons of CO_2, 200,000 tons of SO_2, 80,000 tons of NO_x, and 110,000 tons of particulate matter every year when compared with production of the same amount of electricity from conventional coal-fired plants.

Geothermal power plants are very reliable when compared to conventional power plants. For example, new steam plants at the Geysers are operable more than 99% of the time. Taken as a group, geothermal power plants have system availabilities of 95% or higher. In addition, the capacity factor of geothermal power plants is highest among all types of power plants. *Capacity factor* is the amount of energy actually produced per year in kWh compared with the amount that could be produced if the plant operated continuously at full capacity.

In some parts of the world, geothermal systems are cost competitive with conventional energy sources. It is anticipated that as technology improves, the cost of generating geothermal energy will decrease. Today's cost of electricity from typical geothermal systems ranges from $0.05 to $0.08/kWh. Because they are abundant in the United States, geothermal resources offer a large source of secure energy to the nation's energy portfolio.

The future of geothermal power

Before geothermal electricity is considered a key element of the U.S. energy infrastructure, it must become cost competitive with traditional forms of energy. Toward that end, the geothermal industry, with assistance from the DOE, is working to achieve a geothermal-energy life-cycle cost of electricity of $0.03/kWh. It is anticipated that costs in this range will result in about 15,000 MW of new capacity installed by U.S. firms within the next decade.

Reserves of hydrothermal energy in the United States are difficult to quantify because much exploration remains to be done. However, the U.S. Geological Survey (USGS) estimates that geothermal energy from identified U.S. hydrothermal resources could supply thousands of megawatts more power than current production. In addition, USGS estimates that five times that amount may be available from undiscovered hydrothermal resources in the United States (Fig. 5–6).

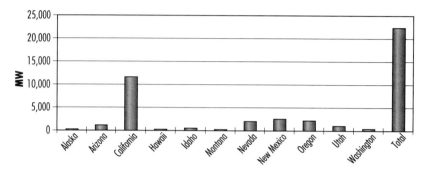

FIG. 5–6 STATE GEOTHERMAL POTENTIAL (SOURCE: GEA AND USGS)

Electricity may also be produced from hot dry rock energy in the future. Hot dry rock energy comes from relatively water-free hot rock found at various depths beneath the earth's surface. One way to access that energy is by circulating water through man-made fractures in the hot rock. Heat can be extracted from the water at the surface for power generation, and the cooled water can then be recycled through the fractures to pick up more heat, creating a closed-loop system. The United States has enough hot dry rock resources to supply a significant fraction of electric power needs, once the technology is developed to make those resources commercially viable.

Geopressured brines and magma may someday add to the electricity mix. Geopressured brines are hot, pressurized, methane-rich waters found in sedimentary basins 10,000–20,000 feet below the surface. *Magma* is molten or partially molten rock within the earth's crust. Current technology has not advanced to the point where geopressured brines and magma can be cost effectively used for energy production.

When hydrothermal reservoirs, hot dry rock, geopressured brines, and magma are considered together, the potential for electricity production from geothermal resources is very large. In fact, with the addition of ambient ground heat used for geothermal heat pumps (not electricity generation), the geothermal resource base in the United States contains about 1.5 million quads of thermal energy. Hydrothermal reservoirs, the only geothermal resources to date that can be economically extracted for electricity production, represent just one-tenth of this resource base.

DOE developments

Federal funding of geothermal research and development is authorized by statute to support the U.S. geothermal industry in providing diversity and therefore security in domestic energy supply options.

This support also helps the industry maintain its technical edge in world energy markets, thereby enhancing exports of U.S. goods and services and U.S. job growth.

Principal research and development thrusts

- economic competitiveness
- reducing geothermal power development costs
- increasing efficiency of geothermal power production
- cutting geothermal maintenance expenses
- environmental benefits
- reducing air and water emissions
- offering alternatives to fossil and nuclear power generation
- producing useful byproducts from geothermal resources
- sustainability
- extending the useful lifetime of geothermal resources
- managing geothermal resources to further reduce net water loss
- limiting plant capacity to match net fluid extraction and natural recharge

DOE is working in partnership with U.S. industry to establish geothermal energy as an economically competitive contributor to the U.S. energy supply.

The long-term viability of geothermal energy lies in developing technology to enable use of the full range of geothermal resources:

- hydrothermal: deep, hot (>90°F) reservoirs with enough energy to generate electricity
- direct use: moderate (68 to 302°F) temperature/depth zones from which heat energy can be directly applied to greenhouses, spas, and other facilities
- shallow depth areas (3 to 4 ft deep): constant-temperature zones supporting geothermal heat pumps to heat and cool residences and other structures

DOE's Geothermal Energy Program is balanced between short-term goals of greater interest to industry, and long-term goals of importance to national energy interests.

DOE goals include:

- increase the number of states with geothermal electric facilities from 8 to 80 by 2006
- reduce the levelized cost of generating geothermal power $0.03 to $0.05 per kWh by 2007
- supply the electrical power or heat energy needs of seven million homes and businesses in the United States by 2010

To achieve these goals, DOE is supporting research in geosciences and supporting technologies, drilling research, and energy systems research and testing. Initiatives include study of 3-D seismic exploration methods, detection and mapping of geothermal fields, and analysis of hydrothermal systems and reservoir dynamics. Drilling research includes near-term improvements to existing technology, improvements in drilling diagnostics and development of innovative drilling subsystems to improve cost-effectiveness.

Research is also ongoing into advanced plant systems and technology, development, and testing of prototype plants to access small-scale field verification. DOE is also backing the *GeoPowering the West* initiative promoting geothermal development in the western United States and the International Clean Energy Initiative, helping U.S. industry identify new international markets in developing countries.

Case Study: New Nevada Campus to be Fully Powered by Geothermal Energy

The University of Nevada, Reno will use geothermal energy to supply all the energy needs for its new Redfield Campus in Reno. Advanced Thermal Systems, Inc. (ATS) will build and operate an 11 MW Kalina Cycle geothermal power plant adjacent to the campus. Under a 30-year agreement with the university, the power plant will provide electricity and hot and chilled water to the university, using an absorption cooling system to produce chilled water from the geothermal heat. ATS plans to sell excess electricity to Sierra Pacific, the local electric utility. The new campus is expected to open in 2004.

ATS affiliates already own a geothermal plant at the southern edge of Reno, near the Redfield campus. The firm hopes to complete the new facility on a timetable complementing first phase construction of the new branch campus.

> *The Kalina Cycle technology is the most efficient method of zero-emissions, 100% renewable geothermal energy production available today. Heat and electrical power can be provided at a price competitive with power plants fired by natural gas or coal, without the emissions of fossil fuel plants. We applaud the University of Nevada for taking this progressive step in demonstrating the viability and effectiveness of geothermal-based energy in Nevada.*
>
> SHUMAN MOORE, PRESIDENT, ATS

The new central utility plant will be designed to meet an initial campus load of about 500 kilowatts (kw) of electricity. It will be scaled up as demand grows with the campus. The plant will provide building heat through a system of heat exchangers. Cooling will be provided through an absorption chilling system.

> *This will be a complete system for powering, heating and cooling the Redfield Campus facilities, all of it connected to the geothermal resources underground. It will be a model not only for universities, but also for other commercial and industrial facilities that want to take advantage of the abundant geothermal resource in Nevada and around the West.*
>
> MOORE

According to the University's Great Basin Center, there are currently 13 power plants operating at nine geothermal sites in Nevada. They produce about 1.2 million megawatt-hours (MWh) of electricity annually, making Nevada second only to California in installed geothermal capacity.

In 2001, the Nevada Assembly passed a new energy standard for the state that will require 15% of power in Nevada to be produced by renewable energy sources by 2013 (*see* Table 5–2).

TABLE 5–2
NEVADA GEOTHERMAL POWER PLANTS

Operator	Plant Name	Year	Net MW
Oxbow	Beowave	1985	16.0
BPP	Brady Hot Springs	1992	21.1
Cal Energy	Desert Peak	1985	8.7
Oxbow	Dixie Valley	1988	66.0
Empire Geo.	Empire	1987	3.6
OESI	Soda Lake 1	1987	3.6
OESI	Soda Lake 2	1991	13.0
Far West	Steamboat 1	1986	6.0
Far West	Steamboat 1A	1988	1.1
Far West	Steamboat 2	1992	14.0
Far West	Steamboat 3	1992	14.0
OESI/CON	Stillwater 1	1989	13.0
Tad's	Wabuska 1	1984	0.5
Tad's	Wabuska 2	1987	0.7
Caithness S.	Steamboat Hills	1988	14.4

Source: DOE

According to the ATS officials, the Kalina Cycle technology will use geothermal heat to vaporize a benign ammonia-water *working fluid*, with vapor produced in the process to drive the electricity-producing turbine generators. The Kalina Cycle technology differs from traditional binary cycle geothermal plants in that it allows for a more efficient delivery of heat through variable vaporization and condensation of the working fluid.

The use of renewable energy and energy efficiency technologies should accelerate in Nevada, thanks to the Nevada Renewable Energy and Energy Conservation Task Force.

The task force, established in late 2001 to administer the state's Trust Fund for Renewable Energy and Energy Conservation, released its first legislative report in 2003. According to that report, the task force plans to work with the Nevada State Energy Office to improve renewable energy resource assessment and examine solutions to power transmission constraints within the state. The task force will also examine credit-trading systems for renewable power producers, evaluate market incentives, consider new energy codes for buildings, and expand the state's public outreach efforts.

Case Study: Calpine and the Geysers

The world's largest geothermal operation is the Geysers, about 100 miles north of San Francisco in the Mayacamas Mountains. The Geysers currently produce about 1300 MW of power. Calpine owns 19 plants at the Geysers, where wells, some greater than two miles deep, have been drilled to tap natural steam (*see* Table 5–3).

TABLE 5–3
GEYSERS GEOTHERMAL AREA

Operator	Plant	Gross MW
Calpine	Unit 1	12
Calpine	Unit 2	14
Calpine	Unit 3	28
Calpine	Unit 4	28
Calpine	Unit 5	55
Calpine	Unit 6	55
Calpine	Unit 7	55
Calpine	Unit 8	55
Calpine	Unit 9	55
Calpine	Unit 10	55
Calpine	Unit 11	110
Calpine	Unit 12	110
Calpine	Unit 13	138
Calpine	Unit 14	114
Calpine	Unit 15	62
Calpine	Unit 16	119
Calpine	Unit 17	119
Calpine	Unit 18	119
Calpine	Unit 19	119
Calpine	Unit 20	119
NCPA	NCPA 1	2 x 55
NCPA	NCPA 2	2 x 55
SMUD	SMUDGEO	78
Santa Fe	Santa Fe	2 x 48
California/DWR	Bottle Rock	55
SMUD	CCPA	2 x 66
Calpine	Bear Canyon	2 x 11
Calpine	Ford Flat	2 x 17
Calpine	Aidlin	12.5
Total		**2,190.5**
Source: DOE		

At the Geysers, the captured steam is piped to generating units. The steam spins a turbine that driving a generator to produce much-needed electricity for the California power market.

Geothermal power at the Geysers has environmental advantages, such as helping to stretch fossil fuel supplies. Presently, this electric-generating system is practical in only a few parts of the world, but innovative technologies are striving to expand and extend the value of geothermal power.

Geothermal outlook

Geothermal resources occur throughout the world, and are particularly abundant in many developing countries where their use could displace the development of polluting fossil fuel-fired power plants.

Geothermal power plants are available in sizes ranging from about 300 kW net electrical output to 55 MW. They can be installed in modules as electrical demands grow. Although not generally recognized, geothermal power generation can and will play a significant part in rural electrification of developing countries, according to statements from GEO.org.

Currently, about 8000 MW of geothermal electrical power generation are on-line in more than 20 countries, including (but not limited to) China, Costa Rica, El Salvador, Iceland, Indonesia, Italy, Japan, Kenya, Mexico, New Zealand, Nicaragua, Philippines, Romania, Russia, Thailand, Turkey, and the United States (*see* Table 5–4).

TABLE 5–4
INSTALLED GEOTHERMAL GENERATING CAPACITY BY COUNTRY

Country	1990 MWe	2000 MWe	Country	1990 MWe	2000 MWe
Argentina	0.67	0	Kenya	45	45
Australia	0	0.17	Mexico	700	755
China	19.2	29.17	New Zealand	283.2	437
Costa Rica	0	142.5	Nicaragua	35	70
El Salvador	95	161	Philippines	891	1,909
Ethiopia	0	8.52	Portugal, Azores	3	16
France, Guadeloupe	4.2	4.2	Russia (Kamchatka)	11	23
Guatemala	0	33.4	Thailand	0.3	0.3
Iceland	44.6	170	Turkey	20.6	20.4
Indonesia	144.75	589.5	USA	2,774.6	2,228
Italy	545	785			
Japan	214.6	546.9	**Total**	**5,831.72**	**7,974.06**

Electricity was first generated from geothermal water at Larderello in Tuscany, Italy in 1904. A 250 kW power station was put into operation at that site in 1913, and the size of the operations grew to 130 MW by the time of their destruction during World War II. After the war, the field was quickly put back into production and it remains an active power generation center.

At Wairakei, North Island, New Zealand, electrical power has been generated from a geothermal resource since 1958, and at the Geysers field, California, USA, power generation began in 1960. Each of these fields is currently operating, and projections indicate they will generate power at commercial rates for many more decades, perhaps centuries.

By some estimates, as much as 80,000 MW of geothermally generated electrical power could be extracted from volcanic systems in developing countries throughout the world. Indonesia alone estimates its potential at 19,000 MW (Fig. 5–7).

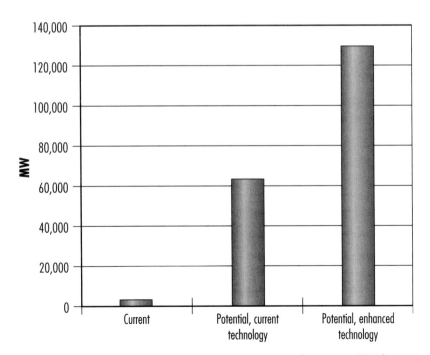

Fig. 5–7 World Geothermal Potential (source: GEA)

Direct applications of geothermal resources are also important, and comprise more than 12,000 thermal MW of use.

For example, the vast majority of the homes and buildings in Iceland are geothermally heated, and district heating systems also used in the Paris basin, in several U.S. cities, in New Zealand, and elsewhere. Many countries across the globe house greenhouses that are heated with geothermal waters. Another established, commercial use of geothermal heat is food drying. The vast potential for direct applications of geothermal resources has barely been touched.

6 Hydroelectric

Since the time of ancient Egypt, people have used the mechanical energy in flowing water to operate machinery and grind grain and corn. However, hydropower had a greater influence on people's lives during the 20th century than at any other time in history. Hydropower played a major role in making the wonders of electricity a part of everyday life and helped spur industrial development. Hydropower currently produces almost one-fourth of the world's electricity. It supplies more than one billion people with power (Fig. 6–1).

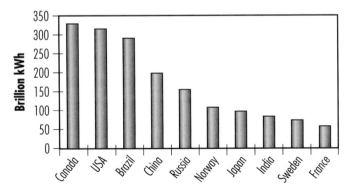

Fig. 6–1 Top Hydroelectric Generating Countries (source: EIA)

The first hydroelectric power plant was built in 1882 in Appleton, Wisconsin to provide 12.5 kW to light two paper mills and a home. Today's hydropower plants range in size from very small to very large, and a few mammoth plants have capacities up to 10,000 MW, supplying electricity to millions of people.

Worldwide, hydropower plants have a combined capacity of around 700,000 MW and annually produce more than 2.3 trillion kWh of electricity (Fig. 6–2).

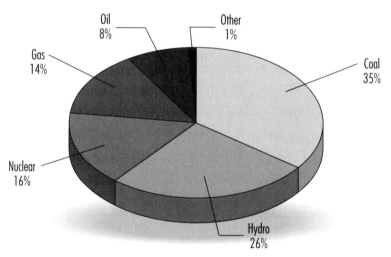

FIG. 6–2 WORLD ELECTRICITY SUPPLY BY SOURCE (SOURCE: EIA)

With a capacity exceeding 92,000 MW—enough electricity to meet the energy needs of 28 million households—the United States is a close second to Canada as a world hydropower producer. Hydropower in the United States supplies about 10% of the country's net generation and accounts for a grand majority of the renewable energy used. When narrowed to utility net generation, hydroelectricity accounts for 7% of the national generation and more than 99% of all utility renewable generation (Figs. 6–3 and 6–4).

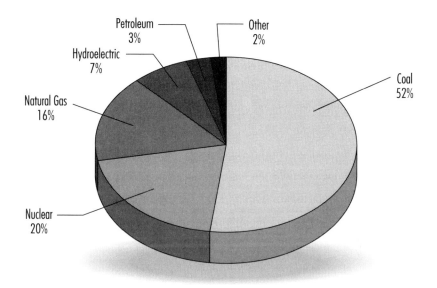

FIG. 6–3 U.S. ELECTRIC UTILITY NET GENERATION BY SOURCE (SOURCE: EIA)

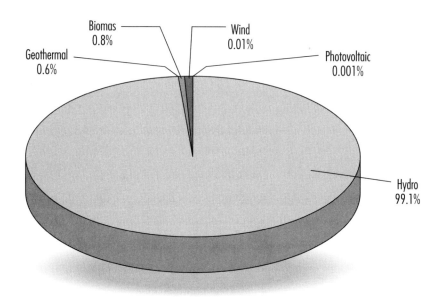

FIG. 6–4 U.S. UTILITY NET GENERATION BY RENEWABLES (SOURCE: EIA)

The nation's largest hydropower plant is the 7600 MW Grand Coulee power station on the Columbia River in Washington State. The plant is second only to the colossal 13,320 MW plant in Brazil.

The renewable energy share of total world energy consumption is expected to decline slightly, from 9% in 1999 to 8% in 2020, despite a projected 53% increase in consumption of hydroelectricity and other renewable resources.

Through 2020, worldwide consumption of renewable energy is projected to increase by 53%, as compared with expected increases of 92% for natural gas and 58% for oil consumption, according to DOE projections. Growth in demand for renewable energy resources is expected to continue to be constrained by relatively moderate fossil fuel prices.

New, large-scale hydroelectric installations are expected to provide much of the growth in renewable energy use in the developing world. China, India, Malaysia, and other developing Asian countries continue to construct or plan large-scale hydropower projects. Construction on the largest project, China's 18,200 MW Three Gorges Dam, continued in 2001 despite reports of corruption and problems in the relocation of populations from the reservoir site. Malaysia continues to work on its 2400 MW Bakun hydroelectric project, although progress has been slow.

The heavy reliance on hydroelectric power in many countries of Central and South America has become a burden for some, because drought has endangered the reliable supply of electricity. In Brazil, persistent drought in 2001 led to a substantial decline in reservoir levels and, therefore, the ability of hydroelectric power plants to provide electricity. Brazil's government enforced a 20% cut-in power use as part of a rationing program, and considered other measures such as reducing the workweek, in an effort to avoid blackouts. In the fall of 2001, reservoir levels were 28% below capacity in key regions of the country. Brazil responded by increasing the pace of natural gas-fired power plant construction, a trend many governments in the region see as necessary in order to diversify electricity supply sources and avoid shortages in the future.

In the industrialized world, Canada is among the only countries with plans to expand large-scale hydroelectric resources, such as the 2000 MW Lower Churchill Project at Gull Island in Newfoundland Province. Many developed countries have already substantially exploited their hydroelectric resources, and increments to their renewable energy consumption are expected to come from wind, solar, and other nonhydroelectric renewable energy sources.

The DOE projections for hydroelectricity and other renewable energy sources include only on-grid renewables. Although noncommercial fuels from plant and animal sources are an important source of energy, particularly in the developing world, comprehensive data on the use of noncommercial fuels are not available and, as a result cannot be included in the projections. Moreover, dispersed renewables (renewable energy consumed on the site of its production, such as solar panels used to heat water) are not included in the projections, because there are also few comprehensive sources of international data on their use.

North America

Hydroelectricity remains the predominant form of renewable energy use in North America, particularly in Canada. In 1999, hydroelectric power provided nearly 60% of the Canada's 551 billion kWh of electricity generation, compared with 8% to 10% in the United States and 14% in Mexico.

Although Canada has announced some plans to expand its hydroelectric capacity over the next decade, hydropower consumption is expected to remain flat or decline slightly over the next two decades in North America. Increases are expected for geothermal, wind, solar, biomass, and MSW.

United States

Potential sites for hydroelectric dams have already been largely established in the United States, and environmental concerns are expected to prevent the development of any new sites in the future. U.S. conventional hydroelectric generation is expected to decline from 316 billion kWh to 304 billion kWh in 2020 as increasing environmental and other competing needs reduce the productivity of generation from existing hydroelectric capacity.

Canada

Currently, 60% of Canada's total installed electricity generation capacity consists of hydroelectric dams. Canada is exploring ways to increase its hydroelectric capacity still further with several proposals that are currently under consideration (Fig. 6–5).

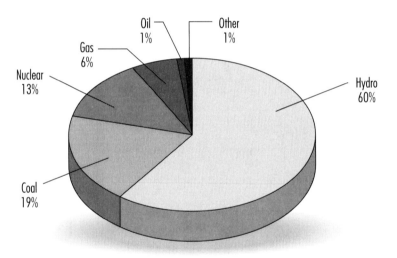

Fig. 6–5 Canadian Electricity Supply by Source
(source: Natural Resources Canada)

In the Northwest Territories there are proposals to develop hydroelectric projects that would total between 12,000 MW and 15,000 MW. The projects would cost an estimated $17.5 billion and would be constructed in a sparsely populated part of the country on six separate rivers: the Mackenzie, Bear, Lockhart, Talston, Snare, and Lac la Marte.

The government has identified 10,000 MW of potential development that could be exploited by developing sites on the Mackenzie River. The largest site, Ramparts, has a potential for 4500 MW. Estimates are that the projects would take between 5 and 20 years to complete.

The successful development of these projects, as well as many others in Canada, will depend on agreements with the local populations that will be displaced or otherwise affected by the projects. In the past, local concerns were not always taken into consideration, and Canadian aboriginal groups began to fight further developments through legal means, often successfully suing developers for reparations or to scale down proposed projects. The current trend is for governments and companies to work with the aboriginal tribes to reach consensus before construction begins, including offers of joint ownership and extensive environmental impact studies. The government of the Northwest Territories is meeting with the indigenous groups affected by hydroelectric development and must reach an agreement with them before any construction begins.

One successful outcome of the new government strategy to gain approval for development from the indigenous people who will be affected by the construction of new hydroelectric infrastructure is the 1200 MW Eastmain Rupert project. In 2000, the provincial utility Hydro-Quebec paid the Grand Council of the Crees some $300,000 to conduct a three-month study of the economic, commercial, and environmental aspects of the utility's proposal to construct the hydroelectric project. The project will cost an estimated $2.5 billion to construct and will involve the diversion of the Rupert River in the James Bay region of Quebec. Although an agreement has been reached between Quebec and the Crees, feasibility studies and environmental authorizations remain to be completed and are expected to take nearly four years. If all approvals are obtained, construction could be completed in 2011.

There are still other plans to construct large-scale hydroelectric projects in Canada. The governments of Newfoundland and Quebec provinces have proposed construction of a 2000 MW Lower Churchill Project at Newfoundland's Gull Island. The project has been scaled back from 2800 MW, because Newfoundland determined it would be too expensive to construct a phase of the project that involved building an 800 MW powerhouse at Muskrat Falls. The government of Newfoundland is studying the Lower Churchill River on Labrador to determine the feasibility of constructing hydroelectric facilities at Gull Island and Muskrat Falls to support proposed aluminum smelters in Newfoundland and Labrador.

Quebec's government has also approved a plan by Hydro-Quebec for construction of a dam and 526 MW powerhouse on the Toulnustouc River, on the north shore of the St. Lawrence River. The project's powerhouse is part of a larger project supported by the Betsiamites Innu-Montagnais aborigines that would include several river diversions. Construction of the 526 MW powerhouse will take an estimated $400 million and will involve enlarging Lake Sainte-Anne reservoir, building a dam and a powerhouse, and connecting the powerhouse to the Micoua substation.

Hydro-Quebec has a number of plans for additional mid-size hydropower projects over the next decade. Quebec's government has authorized Hydro-Quebec to plan a dam and 220 MW powerhouse on the Romaine River near Havre-Saint-Pierre. If approved, construction of the $335 million La Romaine Project could begin in 2004. The station could be commissioned in 2007, generating 1000 GWh annually. Another technical and environmental study has been launched for the development of a 450 MW hydroelectric plant on the Peribonka River nearly 200 miles north of Quebec City. The project would generate an average 2200 GWh of electricity annually. If all goes according to plan, the facility will be ready for commissioning in 2009.

Canadian hydro facts

- Hydroelectricity is the most important source of electricity in Canada.
- More than 60% of the nation's electricity comes from the power of water.
- Hydro accounts for 97% of Canada's renewable electricity generation.
- Every province in Canada, with the exception of Prince Edward Island, has some hydro capacity.
- Provinces that produce the most hydroelectricity are Quebec, British Columbia, Newfoundland and Labrador, Ontario and Manitoba. Together they produce 97% of the country's hydro.
- Canada is the world's second largest exporter of electricity after France. Canada exports more than 45,000 GWh of electricity annually.
- Canada's installed hydro capacity exceeds 67,000 MW, but the country has hydro resources to potentially develop another 118,000 MW.

(Source: Natural Resources Canada)

Central and South America

Hydroelectricity is an important source of electricity generation in Central and South America. In fact, Brazil, the region's largest economy, relies heavily on hydropower, which typically supplies more than 90% of the country's electricity generation. As a result, drought can have a devastating impact on electricity supply, and many countries of Central and South America are initiating projects to diversify the mix of electricity supply.

Much of the diversification will consist of adding natural gas–fired electricity capacity to reduce dependence on hydropower. As a result, although there is some projected growth in the use of hydroelectric and other renewable resources, it is expected to be much less than the growth in natural gas consumption.

Case Study: Energy Crisis in Brazil—Implications for Hydropower

In 2001, Brazil was hit with an energy crisis exposing the risk that accompanies its high level of dependence on hydroelectric power. After the worst drought in 70 years, water levels in many of Brazil's hydroelectric reservoirs fell to critical lows by the summer of 2001. To avert impending blackouts and power interruptions, the Brazilian government introduced a series of emergency measures intended to cut electricity consumption and diversify supply sources.

Industries and commercial businesses were required to reduce their power consumption by 15% to 25%. They were also barred from undertaking any major new expansion works requiring new electricity connections from the main system. Residential use also saw mandated cuts.

Although more than 90% of Brazil's generating capacity and production came from hydroelectric plants, the drought was not the sole factor behind the country's energy crisis. The demand for electricity in Brazil has been growing by almost 5% per year, on average, since 1990. Demand growth has been driven particularly by industrial use in the South, East, and Center West regions, where most of the country's population lives. However, investments in new electricity generation and transmission capacity did not keep pace with demand. The Brazilian government initiated plans to build 49 thermoelectric generators by 2003, fueled primarily by Bolivian natural gas. The absence of power line connections to regions of the South and the North of Brazil, as well as from Argentina, also prevented electricity from reaching the areas facing electricity shortages.

Factors such as private investors' increased perception of risk since the devaluation of the Brazilian real in 1999, the contractual terms of supply offered for natural gas by Petrobras (the federal oil and gas monopoly), and the electricity tariff controls set by Aneel (Brazil's power regulator) are believed to have impeded the capital investments needed to finance new generation and transmission projects in the country.

The slowdown in efforts to privatize the electricity sector in recent years has also contributed to the current energy crisis, because some planned capacity additions were to occur after privatization.

Changes occurring since the drought of 2001 have helped to remove some of the financial and regulatory barriers to electricity sector investment in Brazil. Specifically, the National Development Bank of Brazil made several billion dollars worth of public funds available to companies wishing to enter partnerships in natural gas–fired or hydroelectric power stations. The bank will provide up to 60% of the financing needed by private investors. A formula has also been established to protect investors against exchange rate risk. Furthermore, Petrobras has agreed to a set of supply terms that are considered more favorable by thermoelectric power plant investors, with natural gas to be provided at fixed prices for periods of 12 full months.

On the transmission side, the Inter-American Development Bank approved $243.9 million in financing to build an additional 1000 MW line connecting the electricity grids of Argentina and Brazil.

Substantial governmental effort on the supply side has focused on natural gas–fired generating plants, which can be brought on line faster and at less expense than most other comparable options. Despite the difficulties associated with depleted reservoirs, a significant expansion of Brazil's hydropower infrastructure is also considered a key element of the government's overall plan to shore up the country's electricity supply. Aneel awarded licenses for the construction and operation of about 20 new hydroelectric power plants to add some 5000 MW to Brazil's total generating capacity.

Although expansion of the hydroelectric infrastructure may serve to alleviate electricity shortages in Brazil, it is not without controversy. The development of Brazil's existing hydroelectric facilities has given rise to some economic, social, and environmental problems. The construction of large hydroelectric projects in particular, often beset by long delays and significant cost overruns, has contributed to the country's debt burden since the 1970s.

Reservoir and dam development for large facilities has also disrupted the culture and sources of livelihood for many communities. Studies have indicated that the majority of people uprooted from their existing settlements as a result of dam development are poor and/or members of indigenous populations or vulnerable ethnic minorities. The displaced populations have also had to bear a disproportionate share of the social and environmental costs of large hydroelectric projects without gaining a commensurate share of the economic benefits.

Reservoir and dam development for hydroelectric facilities has also led to loss of forests, wildlife habitats, species populations, aquatic biodiversity, upstream and downstream fisheries, and services provided by downstream flood plains and wetlands.

Hydropower Basics

Hydropower converts the energy in flowing water into electricity. The quantity of electricity generated is determined by the volume of waterflow and the amount of *head*—the height from turbines in the power plant to the water's surface. The greater the flow and head, the more electricity produced.

A typical hydropower plant includes a dam, reservoir, penstocks (pipes), a powerhouse, and an electrical power substation. The dam traps water, which is stored in the reservoir. This also creates the head for the plant.

Penstocks carry water from the reservoir to turbines inside the powerhouse. The water rotates the turbines, which drive generators producing electricity. The electricity is then transmitted to a substation where transformers increase voltage to allow transmission to homes, businesses, and factories.

There are several types of hydropower plants, but two major categories:

Conventional

Most hydropower plants are conventional in design, meaning they use one-way water flow to generate electricity. There are two categories of conventional plants—run-of-river and storage plants.

Run-of-river plants use little, if any, stored water to provide water flow through the turbines. Although some of these plants store a day or week's worth of water, weather changes, particularly seasonal changes, cause run-of-river plants to experience significant fluctuations in power output.

Storage plants have enough storage capacity to offset seasonal fluctuations in water flow and provide a constant supply of electricity throughout the year. Large dams can store several years' worth of water.

Pumped storage

In contrast to conventional hydropower plants, pumped storage plants reuse water. After water initially produces electricity, it flows from the turbines into a lower reservoir located below the dam. During periods of low electricity demand, some of the water is pumped into an upper reservoir and reused during periods of high demand.

In the United States, most hydropower plants are built by federal or local agencies as part of multipurpose projects. In addition to generating electricity, dams and reservoirs provide flood control, water supply, irrigation, transportation, recreation, and refuges for fish and birds. Private utilities also build hydropower plants, although not as many as government agencies (Fig. 6–6).

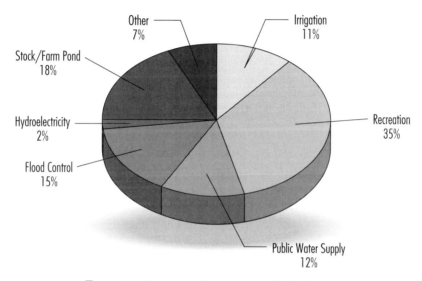

FIG. 6–6 PRIMARY PURPOSE OF U.S. DAMS
(SOURCE: U.S. ARMY CORPS OF ENGINEERS)

Hydropower plants provide inexpensive electricity and produce no pollution. Unlike other energy sources such as fossil fuels, water is not destroyed during the production of electricity—it can be reused for other purposes.

Hydropower plants can significantly impact the surrounding area, however. Reservoirs can cover towns, scenic locations, and farmland, as well as affect fish and wildlife habitat. To mitigate impact on migration patterns and wildlife habitats, dams maintain a steady stream flow and can be designed or retrofitted with fish ladders and *fishways* to help fish migrate upstream to spawn.

The best sites for hydroelectric plants are swift-flowing rivers or streams, mountainous regions, and areas with heavy rainfall. Only 2400 of the nation's 80,000 dams are being used for hydropower, which means many existing dams could be retrofitted to produce electricity, boosting the country's renewable energy base.

Hydropower facilities in the United States can generate enough power to supply 28 million households with electricity, the equivalent of nearly 500 million barrels of oil. The total U.S. hydropower capacity—including pumped storage facilities—is about 95,000 MW. Researchers are working on advanced turbine technologies that will not only help maximize the use of hydropower but also minimize adverse environmental effects.

Types of Hydropower

Impoundment

An impoundment facility, typically a large hydropower system, uses a dam to store river water in a reservoir. The water may be released either to meet changing electricity needs or to maintain a constant reservoir level.

Diversion

A diversion, sometimes called run-of-river, facility channels a portion of a river through a canal or penstock. It may not require the use of a dam.

Storage facilities

When the demand for electricity is low, a pumped storage facility stores energy by pumping water from a lower reservoir to an upper reservoir. During periods of high electrical demand, the water is released back to the lower reservoir to generate electricity.

Sizes of hydropower plants

Facilities range in size from large power plants supplying many consumers with electricity to small and micro plants that individuals operate for their own energy needs or to sell power to utilities. DOE defines hydropower as follows:

- Large: capacity of more than 30 MW
- Small: capacity of 0.1 to 30 MW
- Micro: capacity of up to 100 kW (0.1 MW)

Turbine technologies

There are many types of turbines used for hydropower, and they are chosen based on their particular application and the head available to drive them (*see* Table 6–1). The turning part of the turbine is called the runner. The most common turbines include the following:

Pelton turbine. A Pelton turbine has one or more jets of water impinging on the buckets of a runner that looks like a water wheel. The Pelton turbines are used for high-head sites (50–6000 ft) and can be as large as 200 MW.

Francis turbine. A Francis turbine has a runner with fixed vanes, usually nine or more. The water enters the turbine in a radial direction with respect to the shaft, and is discharged in an axial direction. Francis turbines will operate from 10 ft to 2000 ft of head and can be as large as 800 MW.

Propeller turbine. A propeller has a runner with three to six fixed blades, like a boat propeller. The water passes through the runner and drives the blades. Propeller turbines can operate from 10 ft to 300 ft of head and can be as large as 100 MW.

Kaplan turbine. A Kaplan turbine is a type of propeller turbine in which the pitch of the blades can be changed to improve performance. Kaplan turbines can be as large as 400 MW.

TABLE 6–1
HYDRO TIMELINE

B.C.	Used by the Greeks to turn water wheels for grinding wheat into flour, more than 2000 years ago.	1907	Hydropower provided 15% of electric generating capacity in the United States.
1775	U.S. Army Corps of Engineers founded, with establishment of chief engineer for the Continental Army.	1920	Twenty-five percent of U.S. electrical generation was by hydropower.
		1920	FPA establishes FPC authority to issue licenses for hydro development on public lands.
1880	Michigan's Grand Rapids Electric Light and Power Company, generating electricity by dynamo, belted to a water turbine at the Wolverine Chair Factory, lit up 16 brush-arc lamps.	1933	Tennessee Valley Authority (TVA) established.
		1935	FPC authority extended to all hydroelectric projects built by utilities engaged in interstate commerce.
1881	Niagara Falls, city street lamps powered by hydropower.	1937	Bonneville Dam, first federal dam, begins operation on the Columbia River.
1886	About 45 water-powered electric plants in the United States and Canada.	1937	Bonneville Power Administration established.
1887	San Bernardino, California, first hydroelectric plant in the West.	1940	Forty percent of electrical generation was hydropower.
1889	In the United States, 200 electric plants use waterpower for some or all generation.	Now	About 7% of U.S. electricity comes from hydropower. There is about 80,000 MW of conventional capacity and 18,000 MW of pumped storage.
1901	The first Federal Water Power Act.		
1902	Bureau of Reclamation established.		

Environmental Issues and Mitigation

Current hydropower technology, while essentially emission-free, can have undesirable environmental effects, such as fish injury and mortality from passage through turbines, as well as detrimental effects on the quality of downstream water. A variety of mitigation techniques are in use now, and environmentally friendly turbines are under development.

The goal of the U.S. DOE's Advanced Hydropower Turbine System Program is to develop technology allowing the nation to maximize the use of its hydropower resources while minimizing adverse environmental effects. Conceptual designs of environmentally friendly hydropower turbines have been completed under the DOE-industry program.

Potential injury mechanisms caused by turbine passage have been identified. Research is being performed to understand the effects of these injury mechanisms on fish and to develop methods for reducing their severity.

Potential benefits of advanced turbine technology include the following:

Reduced fish mortality. Advanced turbine technology could reduce fish mortality resulting from turbine passage to less than 2%, in comparison with turbine-passage mortalities of 5% to 10% for the best existing turbines and 30% or greater for some turbines.

Improved compliance with water quality standards. Advanced turbine technology would maintain a downstream dissolved oxygen level of at least 6 mg/L, ensuring compliance with water quality standards.

Reductions in CO_2 emissions. The use of environmentally friendly turbine technology would help reverse the decline in hydroelectric generation and reduce the amounts of CO_2 and other greenhouse gases emitted by consumption of fossil fuels.

Development forecast

The National Hydropower Association (NHA) expects that the existing licensing process will prohibit realizing any new capacity in the future. In fact, NHA is currently predicting a loss of renewable hydroelectric power in the United States without legislative changes to hydropower regulations.

The Federal Energy Regulatory Commission's (FERC) river basin studies show a potential of 73,200 MW of additional U.S. hydroelectric capacity. Emphasizing engineering feasibility and some economic analysis, but no environmental considerations, the FERC estimate is the likely *upper limit of conventional water power potential in the United States.*

The U.S. DOE has undertaken an assessment of hydropower resources using FERC's river basin analysis while also screening for environmental, legal, and institutional constraints at potential sites including threatened or endangered species, national designations, cultural values, and other nonpower issues.

DOE's results show there are 5677 undeveloped hydropower sites with a potential capacity of about 30,000 MW. Of that amount, 57% (17,052 MW) are at sites with some type of existing dam or impoundment, but no power generation. Another 14% (4326 MW) exists at projects that already have hydropower generation, but are not developed to their full potential. Only 8500 MW or 28% of the potential would require new dams.

NHA anticipates that, given the regulatory burden associated with the federal licensing process—the cost, delay and duplication—none of this new capacity will be developed by 2020. Worse, with no changes in the current licensing process, at least 4% of existing hydropower generation undergoing relicensing through 2020 will be lost.

Furthermore, considering the uncertain future of some of the federal projects in the Northwest, the potential loss of generation from our nation's hydroelectric system could be even greater.

However, there are factors that could change NHA's bleak forecast:

- the need for greenhouse gas reductions that would drive domestic policy to again encourage hydropower development
- the hydro licensing process is improved so it increases investor certainty and recognizes the unique energy characteristics and environmental benefit of hydropower
- the resulting licensing rules fairly balance environmental and energy needs

Under these circumstances, NHA forecasts that 20,915 MW of additional power from hydroelectric resources could be developed by 2020—none of which would require the construction of a new dam or impoundment. In terms of greenhouse gas reductions, this would mean displacing 24 million metric tons of carbon emissions from coal.

Hydroelectric generating capacity would rise to 99,478 MW—a 27% increase from current levels—and this nation's use of hydropower resources would rise to 4.9 quads.

Other factors that could further stimulate the development of hydropower capacity include:

- development of commercially viable advanced turbines that further improve biological conditions for fish (fish friendly turbines)
- greater efficiency from these advanced turbines
- the trend in the growing deregulated market to value hydropower's ancillary benefits, its unique ability to stabilize the electric grid
- increased acceptance of green power programs that charge a premium for the delivery of clean and renewable electricity in a deregulated market

Hydro relicensing

Hydro relicensing is a complex and lengthy regulatory process—on average, it takes 10 years to relicense a project, some as long as 20 years. A number of federal laws and regulations, as well as some state laws and regulations, govern the way in which decisions are made and establish the procedures that must be followed. In addition, many governmental agencies are responsible for administering and enforcing these laws and regulations during a relicensing process.

The relicensing of a hydroelectric facility involves a commitment of the nation's resources to electric generation. Accordingly, the public, state, and federal resource agencies have a significant role to play in relicensing decisions. Hydropower, however, is at risk today due to layers upon layers of regulatory requirements and costs associated with the federal relicensing process. Further, hydropower projects on average lose approximately 8% of their generating capacity as a result of the relicensing process.

During relicensing, all relevant environmental, engineering, and legal issues are examined and considered. As a result, relicensing participants may have very different backgrounds, and their interests in relicensing issues are frequently diverse. The process is difficult and at times, contentious. Many of the participants, not only the licensee, have found problems with the process. Because of the complex, expensive and time-consuming nature of the relicensing process, the hydropower industry is presently engaged in several regulatory and legislative arenas to reform the hydro relicensing process.

The need to relicense two-thirds of all nonfederal hydroelectric capacity in the next fifteen years, the momentous changes occurring in the electric power industry due to restructuring, and the increasing need for emissions-free sources of electricity all underscore the need for congressional or administrative action to reform hydropower relicensing.

Under the Federal Power Act (FPA), the FERC has exclusive authority to license nonfederal hydropower projects located on navigable waterways. There are currently about 2500 projects operating under a FERC license. In addition to issuing licenses, FERC also monitors license compliance and dam safety.

A hydropower project license contains terms and conditions that specify how the project may be operated and requires that the project be properly maintained and operated safely. The FPA mandates that FERC issue licenses for a period of 30 to 50 years. Original licenses are typically issued for a 50-year license term. Relicensing is typically issued for a period of 30 to 40 years, depending on the extent of proposed new development or environmental mitigation and enhancement measures. The length of the license term is typically long enough for the owner to recover its economic investment in the project.

When relicensing a project, a hydro owner and FERC not only must comply with the FPA and FERC's regulations but with other laws and regulations. Incorporating the requirements of other laws into the relicensing process is lengthy and complicated. Of these, the National Environmental Policy Act (NEPA) is the most complex and is the primary tool used by FERC to determine if the project should be relicensed.

Legislative and Regulatory Considerations

Since hydroelectric resources have been in use for such a long time, the depth and breadth of regulation and regulatory bodies involved in any hydroelectric project, including any relicensing, is daunting. The following list, courtesy of the NHA, provides a brief overview of the primary legislation regarding hydroelectricity in the United States.

National Environmental Policy Act of 1969 (NEPA)

NEPA requires environmental review whenever a federal action may have significant environmental consequences. In the case of nonfederal hydro projects, FERC's issuance of a license constitutes a federal action. Since a hydro project (which can include dam[s], powerhouse[s], reservoir[s], and surrounding lands) may contain historic structures, discharge waters into receiving streams, and provide habitat for plants, animals, and invertebrates, it is important to assess the effect of project operations on all of these resources.

Clean Water Act of 1977 (CWA)

The CWA prohibits the discharge of pollutants or fill into most waterways of the United States without a permit issued under EPA's National Pollutants Discharge Elimination System (NPDES) or the Corps of Engineers' Section 404 permit. Section 401 of the CWA requires that before FERC issues a license to construct or operate a hydroelectric project, the owner must secure a water quality certificate (*401 Certification*) from the state water quality agency.

Fish and Wildlife Coordination Act of 1934 (FWCA)

The FWCA requires FERC to first consult with the U.S. Department of Interior, Fish, and Wildlife Service and appropriate state fish agencies before granting a license to a hydro owner to control, impound or modify a stream or water body.

National Historic Preservation Act of 1966 (NHPA)

The NHPA requires FERC to take into account the effect of issuing a license for a project on any district, site, building, structure, or object that is included in or eligible for inclusion in the National Register of Historic Places (NRHP). In such cases where there would be an effect, FERC must provide the advisory council on historic preservation the opportunity to comment on the relicensing of a project.

Wild and Scenic Rivers Act of 1968 (WSR Act)

Section 7(a) of the WSR Act (a) prohibits FERC from issuing a license for the construction of any project *on or directly affecting* a wild and scenic river, and (b) limits the power of any federal agency to assist in the construction of any *water resources project having a direct and diverse effect on the values for which the river was established.* The National Park Service is responsible for maintaining a list of all designated rivers under the WSR Act and those that are being studied for inclusion under the WSR Act.

Endangered Species Act of 1973 (ESA)

FERC must consult with the FWS or National Marine Fisheries Service (NMFS) to determine whether the relicensing of a project is likely to jeopardize the continued existence of any endangered or threatened species or result in destruction of critical habitat.

Coastal Zone Management Act of 1972 (CZMA)

FERC must comply with Section 307(c)(3)(A) prior to issuing a license. FERC must certify to the state that the proposed activity (i.e., issuing a license to operate, or continue operating, the project), whether in or outside the coastal zone, affecting any land or water use or natural resource of the coastal zone, will be conducted in a manner that is consistent with the enforceable policies of the state's Coastal Management Program (CMP).

Americans with Disabilities Act of 1990 (ADA)

This act was created to protect the civil rights of persons with disabilities. Titles II and III of the ADA apply to licensee's recreation facilities. This law requires public and private entities that have public accommodations, such as recreation facilities at hydro projects, to be accessible to persons with disabilities. FERC requires new facilities and access areas to comply with the requirements of the ADA, although FERC does not enforce requirements of the ADA. The Department of Justice is responsible for enforcing the ADA.

Case Study: Hydropower in Hawaii

Hawaii has several hydropower plants located on the Island of Hawaii, Kauai, and Maui. Although they are small in comparison to many mainland facilities, they have furnished power to sugar mills and the three island utility companies for many years.

Imported oil is used to supply about 90% of Hawaii's energy needs. No place else in the United States is so critically dependent on imported oil.

Unlike the mainland, Hawaii can't turn to neighboring states to make up for any temporary or permanent energy shortages. Unlike any other state, imported oil is the single thread that can completely unravel Hawaii's future.

Hawaii is blessed with a variety of other energy sources—lots of sunshine, strong winds, fast-growing crops, flowing streams, geothermal heat, and both warm and cold ocean waters. All these resources have the potential to help produce energy and reduce our dependence on imported petroleum.

Hawaii is determined to explore the best ways to take advantage of its renewable energy resources. Each renewable resource helps curb global warming by reducing air pollution. Every barrel of oil or ton of coal replaced with these renewable resources will reduce the amount of CO_2 put into the atmosphere.

The State of Hawaii is actively supporting the development of a mix of renewable energy resources including solar power, biomass, hydropower, wind power, geothermal energy, and ocean thermal energy conversion.

Kauai

Seven hydropower plants, ranging in size from 0.5 MW to 3.8 MW, are located on Kauai. They are operated by Gay & Robinson Sugar Company and AMFAC Sugar Company, furnishing power for sugar plantation and mill operations. Surplus electricity is sold to the Citizens Utilities Company, Kauai Electric Division. In 1996 these hydropower plants supplied 10.4% of Kauai's electricity needs.

Maui

Three hydropower plants are operated by Hawaiian Commercial & Sugar Company and one by Pioneer Mill on Maui. The largest has a capacity of 4.5 MW and is located on the Wailoa Ditch. During 1996 these plants furnished approximately 1.4% of Maui's electrical needs.

Hawaii

Eleven hydropower plants are located on the Island of Hawaii. Five are located on the Wailuku River near Hilo. Four of these are owned and operated by the Hawaii Electric Light Company (HELCO) and range in size from 0.4 MW to 1.5 MW capacity. The fifth hydropower plant located on the Wailuku River is the 12 MW facility operated by Wailuku River Hydroelectric Power Company. The County Department of Water Supply, Wenko Energy Company, or farms operate other plants on the island. They supply power for their own operations and can also provide electricity to HELCO. Hydropower plants provide about 5% of the Island of Hawaii's electrical needs.

The Wailuku River Hydroelectric Power Company plant is the largest in the state and began producing electricity in May 1993. The plant, which took five years and $30 million to plan and build, is located on state conservation land about five miles from Hilo. A diversion channel was dug to feed three miles of five-foot diameter pipe (penstock) with rushing water for the plant's two turbines.

Fuel Cells

Fuel cells are being developed for a variety of applications, including automobiles, cell phones, laptop computers, backup power, residential energy systems, microcombined-heat and power, commercial and industrial base load power, and utility grid support. It is expected that in the next few years, the opportunity for fuel cells could exceed 1000 MW/year on a global basis.

Fuel cells, known in concept for more than 150 years, are poised to make significant contributions to stationary power generation. The initial impetus for development occurred in the late 1950s when the National Aeronautics and Space Administration (NASA) began searching for a compact electricity generator to provide on-board power for manned space missions. Nuclear reactors were too risky, batteries too heavy and short-lived, and solar power was too cumbersome. NASA eventually funded more than 200 research contracts into all aspects of fuel cell technology, ultimately leading to successful applications for the Apollo and Space Shuttle programs.

Approximately 30 years and $1 billion in research have been devoted to addressing the barriers to the use of fuel cells for stationary application. Like batteries, fuel cells are a solid-state device producing power without combustion or rotating machinery. They make electricity by combining hydrogen (H_2) ions, drawn from a hydrogen-rich fuel, with oxygen atoms. However, unlike batteries, which provide the fuel and oxidizer internally, requiring periodic recharging, fuel cells use a supply of these ingredients from outside the system and produce power continuously as long as the fuel supply is maintained.

Fuel cells consist of two electrodes separated by an electrolyte, such as phosphoric acid or molten carbonate. Oxygen passes over one electrode and H_2 passes over the other. As the H_2 is ionized, it loses an electron. The H_2 and the electron take separate paths to the second electrode before combining with the oxygen. The H_2 migrates through the electrolyte and the electron moves through an external circuit.

Since there is no combustion involved, emissions from fuel cells are inherently cleaner than associated emissions from gas turbines, internal combustion engines, or other fossil fuel burning plants. In fact, when pure H_2 is used as fuel, the only effluents are power, heat, and water.

Fuel cells can operate on gasoline, diesel oil, methane, plant matter (bio-gas), or other hydrocarbons as a fuel stock, but the only fuel the cell actually uses is the H_2.

A typical fuel cell plant is actually made of three independent but connected parts. At the user end, there is typically a power conditioner or inverter, which takes voltages and currents produced within the cell and makes them useable for the customer. These power conditioners can be tailor-made for individual customer needs. In the middle is the fuel cell stack itself, which is a collection of individual fuel cells electrically connected in series.

In a fuel cell plant operating with a fuel other than pure H_2, there is some type of fuel processor or reformer on the front end, designed to make high quality H_2 fuel out of a fuel normally too *dirty* for a fuel cell.

One popular method for splitting the H_2 out of fossil fuels is called steam reforming. Steam reforming is a process in which steam and methane (or another fuel) are combined at high temperature and pressure, initiating a chemical reaction creating H_2 and CO_2. The H_2 is then used in the fuel cell. Other methods of separating H_2 include electrolysis and extraction of H_2 from other industrial processes.

Scientific Beginnings: How Fuel Cells Work

The theoretical basis of the fuel cell was explored by Sir William Grove before the American Civil War around 1839. Having discovered that water could be decomposed into its component elements through the application of electricity, Grove proved it was possible to reverse the process to produce an electrical current.

The basic structure of a single fuel cell consists of an electrolyte layer in contact with a porous anode and cathode on either side. A cell produces a small amount of power. Commercial fuel cells are actually a collection of individual cells, called stacks. In these stacks, cells are connected in an electrical series to achieve a voltage that is practical for the given load.

Hydrogen, or a hydrogen-rich fuel, is introduced into the fuel cell at the anode of each cell. Here a reaction occurs in which the H_2 atom splits into a proton and an electron. Freed electrons exit through the external electrical circuit as DC electricity, while H_2 ions (the proton) pass through the electrolyte to the cell's cathode. The flow of electrons returns to the cell stack at the cathode of each cell. In the cathode, returning electrons react with H_2 ions and oxygen from the air to form water.

In theory, any substance that can be chemically oxidized can be supplied continuously (as a fluid) as the fuel at the anode of the fuel cell. The oxidant can be any fluid that can be reduced at a sufficient rate.

Hydrogen is the fuel of choice for most applications, because of its high reactivity when the proper catalyst is used and its ability to be produced from common hydrocarbons.

Oxygen from ambient air is the oxidant of choice for a fuel cell because it is easily and economically obtainable. Air is composed of approximately 21% O_2. Using air, rather than pure O_2, reduces the current density. Polarization at the fuel cell cathode increases with a boost in O_2 levels.

In fuel cells with liquid electrolytes, the reactant gases diffuse through an electrolyte film that *wets* portions of the porous electrode and reacts electrochemically on the electrode surface. Too much liquid electrolyte will *flood* the electrode, restricting the diffusion of gases. This reduces electrochemical performance at the porous electrode. Significant research and development efforts are aimed at thinning fuel cell components while improving the electrode structure and the electrolyte phase. These improvements provide higher and more stable electrochemical performance.

Three Main Fuel Cell Types

A major distinguishing characteristic of different fuel cells is the type of electrolyte used. The three main fuel cells developed for stationary power generation are named for the type of electrolyte used (*see* Table 7–1):

- phosphoric acid fuel cell (PAFC)
- molten carbonate fuel cell (MCFC)
- solid oxide fuel cell (SOFC)

TABLE 7–1
FUEL CELL TECHNOLOGY COMPARISON

Type	Size	Temperature	Efficiency	Advantages	Disadvantages
Phosphoric acid	200 kW	190–210°C	37–45% 75–80% CHP	Commercially available Proven reliability Low CHP potential	External reformer, corrosive electrolyte Large size and weight, low cost cut potential High operating cost Expensive catalyst
Molten carbonate	200 kW to 10 MW	650°C	50–60% 80–85% CHP	Strong CHP potential Fuel flexibility, internal reforming No rare materials Nearing commercialization	Expensive, long start-up cycle Inflexible to load change Heat shield needed, efficiency affected by temperature
Solid oxide	<1 kW to several MW	650–1000°C	45–55% 80–85% CHP	Top CHP potential, high efficiency Fuel flexibility, internal reforming Scalable, potential for load following	High temperature speeds cell breakdown Heat shield needed, high capital cost Efficiency sensitive to temperature
Proton exchange	1–250 kW	80–100°C	30–40% 55–70% CHP	Fast start-up, high power density Low temperature, possible load following Long cell and stack life, potential cost cuts	Lower efficiency and CHP potential High material costs, external reformer Sensitive to fuel impurities

Power Engineering, May 2002, Reed Wasden Research.

PAFC

The most commercially developed type of fuel cell is the PAFC, already installed at hospitals, nursing homes, hotels, office buildings, and utility power plants. PAFCs generate power at more than 40% efficiency and about 85% of the steam produced is used for cogeneration.

Liquid phosphoric acid is the electrolyte used in PAFCs and platinum is the required catalyst for the electrodes. Reforming of the natural gas feedstock to a hydrogen-rich gas occurs outside the fuel cell stacks.

Unless a fuel cell is using pure H_2 gas, the fuel has to be reformed prior to use in the fuel cell. PAFCs require reformation external to the cell. The external reformation is needed because the internal operating temperature of the fuel cell plant is not sufficient to reform the fuel internally. Carbon monoxide must be diluted to below 5% by volume to avoid poisoning the platinum catalyst.

United Technology subsidiary, ONSI Corporation, has more than 100 operating PAFCs worldwide. ONSI's PC 25 fuel cell plant is a 200 kW standalone unit that measures 18 ft x 10 ft x 10 ft. Its compact size and quiet operation (conversation is possible at normal volume levels right beside the unit) allow the unit to be sited with the load, whether that load is a communications center or a university library.

The footprint of this generator, at 180 square feet, is rather large in terms of square feet per kW. Diesel generators and gas turbines provide a much higher number of kW per square foot than the PC 25. However, combustion generators often require sound-dampening enclosures or buildings, which enlarge their footprints. Even with enclosures, combustion generators are often too noisy to locate immediately adjacent to their loads.

PAFCs are not ideal candidates for emergency power. PAFC owners tend to use them for baseload or primary power, tapping into the grid for emergency backup and peak loads. Typical PAFC start-up time is three hours.

Due to the low operating temperature of PAFCs (as well as PEMs and AFCs), noble metal catalysts, such as platinum, are needed to produce adequate reactions at the anode and the cathode. Hydrogen is the only acceptable fuel for use in these fuel cells. Fuels from which H_2 can be derived must be reformed before the fuel can be used in a low temperature fuel cell. Carbon monoxide contaminates or *poisons* the anode in low temperature fuel cells, but can be oxidized in high temperature fuel cells using metals such as nickel for an electrocatalyst.

MCFC

MCFCs eliminate external fuel processors. Methane (the main component of natural gas) and steam are converted into a hydrogen-rich gas in the reforming anode or in a reforming chamber, which are part of the fuel stack.

Using less expensive nickel-based electrodes, MCFCs' electrolyte is molten carbonate salt. When heated to 650°C, the salts melt and provide ionic conductivity. At the anode, H_2 reacts with the carbonate ion from the electrolyte to produce water, CO_2, and electrons.

At the cathode, oxygen from processed air and CO_2, recycled from the anode, react with electrons to form carbonate ions, which then replenish the electrolyte and transfer current through the fuel cell.

MCFCs are expected to deliver fuel-to-electricity efficiencies of 50% to 60%, independent of plant size. Because of the high operating temperatures, MCFCs are more flexible in the type of fuel used.

The higher temperature also allows easier and less complicated fuel reformation. Reforming can take place inside the plant, resulting in a significant efficiency increase. Carbon monoxide is a directly usable fuel in MCFC plants. Reactions occur with nickel catalysts, rather than the more expensive precious metals. MCFCs are expected to be able to use gasoline, diesel, and coal-based fuels such as gasified coal. This fuel requires stringent cleanup to meet the requirements of a fuel cell, however, which is a difficult and costly process. Also, heat from MCFCs can be used to drive a gas turbine or to produce steam for use in a steam turbine for cogeneration.

Carbon dioxide is produced by fuel cells using carbon-based fuels. While CO_2 is a greenhouse gas, the CO_2 produced in a fuel cell reaction is a mere fraction of the pollutant-to-power ratio in current fossil fuel combustion processes. In MCFCs, CO_2 is used in the cathode reaction to maintain carbonate concentration in the electrolyte. So CO_2 is produced in the anode, and consumed at the cathode.

SOFC

SOFCs use a ceramic, solid-state electrolyte. The oxide ions, transported through the electrolyte, combine with H_2 at the anode and produce water vapor and release electrons to the external circuit. At the cathode, oxygen combines with electrons to produce oxide ions. Carbon monoxide can be used instead of H_2 to produce CO_2 and electrons in the anode.

Figure 7–1 shows the world's first fuel cell/gas turbine power plant. The system combines the Siemens Westinghouse SOFC with an Ingersoll Rand microturbine at the National Fuel Cell Research Center on the campus of the University of California at Irvine. Siemens Westinghouse and Southern California Edison announced that the hybrid has passed the major endurance phase of testing. The integrated SOFC/microturbine hybrid produces about 200 kW.

FIG. 7–1 WORLD'S FIRST FUEL CELL/GAS TURBINE POWER PLANT GENERATES ELECTRICITY
(PHOTO COURTESY OF SIEMENS WESTINGHOUSE)

SOFCs are attractive for utility and industrial applications due to several features. They have high tolerance to fuel contaminants, and the high temperature of reactions does not require expensive catalysts, allowing direct fuel processing. Because the electrolyte is solid, problems such as electrolyte flooding and electrolyte migration are avoided.

SOFCs promise efficiencies of 60% in large applications such as mid-sized power generating stations and large industrial plants. This type of fuel cell uses a ceramic material, rather than a liquid electrolyte, because the ceramic lends itself well to the high operating temperatures (1000°C) and fuel flexibility expected in a SOFC.

Similar to the MCFC, CO is a usable fuel for an SOFC. Unlike an MCFC, however, CO_2 is not required at the cathode. The high operating temperature allows for internal fuel reforming with the addition of reforming catalysts. Again, the high temperatures make this fuel cell plant an ideal candidate for cogeneration applications.

Future applications of the SOFC power plant are expected to be linked to a gas turbine in a combined-cycle application with efficiencies as high as 85% when waste heat from the process is used.

The high operating temperature has some drawbacks, however. Thermal expansion mismatches among materials and seals between cells are difficult in flat-plate configurations. Design changes, such as casting the cell into different shapes, may help alleviate this problem.

Polymer exchange membrane (PEM) and alkaline

Two other fuel cell types developing for commercial applications are PEM and alkaline. PEMs have high power density, can vary their output quickly to meet shifting power demands and are well suited to applications where quick start-up is needed, such as in automobiles. PEM cells are targeted for use in continuous premium-power service and as small peaking generators in retail markets. Figure 7–2 shows a worker inspecting a PEM membrane assembly, the heart of the PEM fuel cell technology.

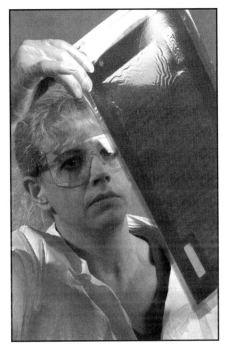

FIG. 7–2 PEM MEMBRANE ASSEMBLY
(PHOTO COURTESY OF UTC FUEL CELLS)

A plastic sheet called a membrane separates the anode and cathode in PEMs. The membrane acts as a one-way valve, blocking electrons, H_2 gas, and oxygen gas generated at the anode while allowing positively charged protons from the anode to diffuse through the membrane to react with negatively charged oxygen radical to form water. The cathode expels the water and adsorbs more oxygen to start the initial reaction all over again.

Alkaline fuel cells have long been used on NASA missions and they achieve power generating efficiencies of 70%. Alkaline potassium hydroxide is used as the electrolyte. Each NASA fuel cell is capable of providing 12 kW continuously and up to 16 kW for short periods.

Polymer electrolyte fuel cells (PEFC), or proton exchange membranes (PEM), function without the addition of an electrolyte, other than the membrane itself. The cell consists of a proton-conducting membrane sandwiched between two platinum porous electrodes. The electrochemical reaction is similar to that of the PEFC.

Because of the characteristics of this type of cell, a low operating temperature of about 80°C is possible. The PEM is also able to sustain operation at very high current densities. This leads to a fast-start capability, ideal for many applications. Since there is no liquid electrolyte in a PEM, its orientation is irrelevant to its power producing capability. This characteristic makes the PEM an ideal candidate for automotive power applications.

There are drawbacks of the relatively low operating temperature of the PEFC. One is the inability to use the reaction heat in cogeneration mode. Because there is little heat produced, endothermic reforming must take place external to the fuel cell. Carbon monoxide is a poison to the platinum plates in a PEFC at this low temperature. Even reformed hydrocarbons contain about 1% CO. Additional CO removal is required because the PEFC is sensitive to CO even to the low ppm level.

Strength in diversity

Current technology includes several different types of fuel cells, each with unique characteristics and application. Fuel cell types are characterized by their electrolyte, though the operation is fundamentally the same from one type to another.

The four major categories of fuel cells currently in use for development of transportation and power generation are:

- phosphoric acid
- molten carbonate
- solid oxide
- polymer electrolyte (proton exchange) membrane

Race to commercialization

Cost per kW is the primary factor holding back fuel cell commercialization. Their cost is about double that of other fossil fuel power systems. Cells carry several important advantages over other fossil fuel systems, however, including a lack of harmful environmental effects, fuel flexibility, low noise generation, and high efficiencies.

Fuel cell researchers are working to bring down the initial costs of fuel cells through the development of less expensive components and several other strategies. Some researchers are working to beat the cost hurdle by boosting efficiencies.

One of the main things needed to get costs down is the elusive economic gains of mass production. Inroads are being made though, particularly through specialized installations where the initial cost is less of a factor than suitability.

Companies are also working to increase the fuel flexibility aspect of fuel cells, which gives them added value. Even for technologies developed to use a variety of fuels, it is necessary to find near-term niche applications that can tolerate the higher initial investment required.

Combined heat and power applications, which raise the efficiency of fuel cells, are a promising area for the technology. Plug Power has a CHP system, called the GenSys5C. The first unit is installed for the Long Island Power Authority (LIPA) at the town hall in Babylon, New York, where it provides supplemental heat and power. The cell is interconnected to the LIPA power grid (Fig. 7–3).

FIG. 7–3 PLUS POWER'S GENSYS5C CHP PACKAGE
(PHOTO COURTESY OF PLUG POWER)

The system, based on Plug's proprietary technology, captures heat generated during the production of electricity and makes it available for integration into a heating or hot water system. The system is capable of generating 5 kW of power and 9 kW of heat. The units are available for purchase with a lead-time of 8 to 10 weeks. To develop the system, Plug placed a number of different systems in the field and watched them run, learning from their performance. The company then took its findings and developed the integrated system. Plug is still working to advance and improve the system, with an eye toward reaching commercial markets in Europe and Asia as they develop, followed by installations in the United States.

Gaining popularity

Fuel cells are currently being developed for a multitude of applications, including automobiles, cell phones, laptop computers, residential energy systems, micro CHP and baseload power. By 2005, the opportunity for fuel cells could top 1000 MW annually on a global basis, according to Siemens. At $1500/kW, this amounts to $1.5 billion a year.

That's a rosy potential, but for the most part, fuel cell companies are still several years away—at least—from turning a profit.

Reed Wasden Research, generally bullish on fuel cells, believes companies could be profitable by 2006 and poised to capture a significant share of the $4 billion to $10 billion distributed generation market by 2010. Reed Wasden tempers the optimism by saying, "Predicting profitability for fuel cells companies is not a precise science and might not occur within the decade."

Capital costs present the biggest challenge for fuel cells, especially when compared to other power generation technologies such as reciprocating engines and gas turbines. Commercially available phosphoric acid fuel cells sell for about $4500/kW, but the manufactured cost of the other fuel cell types are still about three to four times higher.

> *Fuel cells are not designed to compete with a 500 MW combined-cycle power plant. We're competing with the cost of electricity the utility can deliver to the customer's door.*
>
> JERRY LEITMENT,
> PRESIDENT AND CEO, FUELCELL ENERGY (FCE)

In areas where congestion prevents the cost-effective addition of expensive transmission and distribution upgrades, or in areas where environmental concerns preclude, or markedly raise the price of installing other more traditional types of generation, fuel fells can be economic winners.

Matching the $500/kW to $600/kW capital cost for a combined-cycle gas turbine installation is not necessary in these scenarios.

FCE is working toward a manufacturing capacity of 400 MW/yr with an installed capital cost for its direct fuel cell plant in the $1500 to $1800/kW range. At those levels, assuming a 20-year life for the plant on a five-year stack replacement schedule, and assuming $4/MMBtu gas and a 10% to 12% cost of money, the levelized cost of electricity at the fuel cell installation equals $.05 to $.06/kWh—a competitive level at the point of use for many consumers.

> *We see fuel cells as complementary to central station power, not as a replacement for central station power.*
>
> ALLAN CASANOVA,
> FUEL CELL DIVISION, SIEMENS WESTINGHOUSE

Siemens Westinghouse is developing SOFC systems for applications from 5kW to 10 MW. The group is also building a manufacturing facility to produce up to 100 MW of fuel cells annually in 2006.

Siemens Westinghouse estimates electricity production costs from current models runs about $.15/kWh, which makes the fuel cells viable only for niche applications. The firm is striving to reduce capital costs to drop production costs to the $.06 to $.10/kWh range, where energy security and power quality can make the cells a more attractive choice.

Fuel cells themselves still offer some room for technological improvements to drive capital costs down, but the greatest cost cuts may come from the economies of scale gained through commercialization. FCE and Siemens Westinghouse are moving efforts out of the laboratories and into factories to capitalize on higher-volume manufacturing techniques, including automation, supply chain management and production standardization. Balance-of-plant materials account for about 30% of total cost and they may be driven lower through higher volumes and standardization.

Not everyone is convinced that the *build it and they will come* philosophy is correct. UTC Fuel Cells is changing its approach to commercialization for its new line of PEM fuel cells, contracting with high-volume suppliers and following the model in the automotive industry to receive the cost reductions needed to get its planned 150 kW product down to $1500 kW.

Several PEM fuel cells companies are currently developing productions, including Plug Power, H Power, DCH Technology, Ballard Power, Proton Energy Systems, and others; a variety of strategies and commercialization models are likely to emerge.

No one is really sure which are the best size niches, although the 200–300 kW range is gaining interest due to its potential fit with commercial and industrial cogeneration applications. FCE intends to manufacture and market 300 kW, 1.5 MW, and 3.0 MW fuel cell systems. FCE believes these sizes represent a diverse mix of sizes to meet the needs of various commercial and industrial customers. FuelCell intends to get its systems out before its competitors to capture first mover status.

For residential systems, 5 kW is a commonly accepted, likely size target, but that is subject to change depending on residential electricity consumption trends and the demographic profile of potential buyers. PEM fuel cells appear to be the leading technology candidates for residential use because of their low-temperature option and high power density. Siemens has been awarded a contract by the U.S. DOE as part of the Solid-State Energy Conversion Alliance to develop and advance SOFC technology with the focus on a $400/kW target by the end of the decade. The goal of this program is to develop low-cost systems at 5 kW.

Reliability

Despite their long-proven use on space missions, it's important to remember that fuel cells are relatively unproven in the applications for which they are currently being designed. The longest running fuel cells for power generation or cogeneration applications have at most 20,000 hours of operation, far less than the intended 40,000 to 100,000 hour projected lifetimes.

Even if mass production or automotive-like assembly operations bring capital costs down to $1500/kW or lower, long-term reliability remains a relative unknown. For the most part, however, fuel cell manufacturers are confident of their ability to deliver reliable products. Siemens points to demonstration units with 20,000 hours of operation in which stack degradation has been less than 0.1% for every 1000 hours of operation. UTC Fuel Cells expects to incorporate into its PEM fuel cell products the lessons learned from its commercial PAFC units, many of which have operated for five years without any durability problems. FCE expects its units to meet the industry target of 40,000 hours of operating based on extrapolation of the decay rate demonstrated in test plants, which have 0.25% decay per 1000 hours of operation, or 10% loss of power over the 40,000-hour life.

The hydrogen economy

Some people see the evolution of the fuel cell industry as the first step in the transformation to a global H_2 economy, a concept that has been discussed for decades without significant progress. Purposely or not, the fuel cell developers may be beginning the charge in this direction, but to build up necessary momentum, more muscle may be needed.

The conventional fuel suppliers in the current fossil fuel based economy—Exxon-Mobil, BP, Royal Dutch Shell, ChevronTexaco, and the other oil majors and supermajors—will have to buy into the concept in order to build the distribution systems needed. The oil majors already possess the expertise needed to manage the conversion to an alternative fuel infrastructure.

The automakers could be another huge factor. The automobile industry has already put billions of dollars into fuel cell development. To date, the investment hasn't paid off, because the efficiency and environmental advantages of fuel cell vehicles are not enough to overcome the fuel cell's current economic disadvantage when compared to the internal combustion engine.

Fuel cell capital costs need to fall all the way to the $20 to $30/kW range to be competitive with internal combustion engines. In the meantime, automakers may work to recoup some of their investment by pushing fuel cell products into the stationary power markets. Automakers have significant advantages, including deep pockets, massive marketing arms, mass production savvy and brand awareness. Capitalizing on these advantages, automakers may be the best-equipped industry to accelerate the broad implementation of fuel cells.

Residential applications are probably much farther from commercialization than larger stationary power applications due to current high costs and the proven incumbent technology. The automotive industry's advances and funding may help speed this evolution, but widespread availability is not likely before then end of this decade.

Geography

North America has great interest in fuel cell technology, but its initial widespread use may well come in other parts of the world. Mosaic Energy, for example, is relocating its commercialization efforts, including its initial product PEM offerings and market entry, to the Pacific Rim. Mosaic Energy is still interested in the U.S. market and its enormous potential, but management sees the U.S. market as a later opportunity. Demonstration and commercialization will come first in Japan, with the United States to follow.

Europe may be an earlier adopter of fuel cells as well. Environmental regulations, including CO_2 taxes, are stricter in most European countries than in the United States. Europeans in general are typically more environmentally aware than their American counterparts. Furthermore, since the prices for most energy commodities in Europe are significantly higher than those in the United States, the introduction and acceptance of fuel cell technology may have an inherent advantage.

Late in 2002, the EU announced its commitment to achieving a H_2 energy infrastructure, vowing to overtake the United States and Japan in the race to a hydrogen-based future.

Leading a European drive toward a massive increase in H_2 research and development, the European commission president announced that the scientific program will be as important for Europe as the space program was for the United States in the 1960s, but with an economic payback far greater. If successful, H_2 power also would relieve Europe from a potentially dangerous and growing reliance on oil and gas imports, and address the concerns of the region's politically powerful green lobbies.

The EU plans to spend $2.09 billion between 2003 and 2006 on renewable energy development, mostly on technologies related to H_2. That's up from $124 million spent between 1999 and 2002. Hydrogen converted to electricity by fuel cells could power vehicles and provide stationary power for almost any application.

According to heads of the EU, H_2 power, although still years from widespread use, has reached a point where it presents a realistic alternative to fossil fuels. Government financial support and legislation can now push the technology toward practical use, which would push Europe into the global lead and triggering a wave of scientific achievement.

The European initiative followed U.S. debates on its energy future. The U.S. Congress has considered proposals to require utilities to supply up to 10% of their power from renewable energy sources, and some states have already enacted mandates.

The EU committed itself to generating 22% of its gross electricity consumed through renewable sources by 2010 with 12% of all energy coming from renewable sources by that same date.

The EU leaders compared the importance of the H_2 initiatives with the introduction of the euro and EU enlargement, saying that the technology carries a higher priority in Europe than in the United States where fuel is cheaper.

Although the U.S. public and private sectors had been spending more on fuel cell research than the EU, European leaders contend that the U.S. investments aren't well coordinated and therefore won't be as effective as Europe's program.

Incentives can provide another push for new technology. Germany, for example, offers a 5.11 eurocents/kWh credit for 10 years for grid-connected combined heat and power fuel cell systems up to 2 MW. Japan offers a 50% cost share program for fuel cell projects, particularly aimed at wastewater applications.

Incentive programs are emerging in the United States as well, but funding levels are often restricted, and working through the red tape to access the funds can often impose significant delays on fuel cell projects. Incentive programs could accelerate market penetration. To get beyond niche markets to establish fuel cells in a broad sense, incentives are needed. Sales volume is needed to drive costs down, but the sales volume won't be achievable until costs are reduced. However, incentives can help bridge the gap.

Another area that needs a bridge is the relationship between fuel cell developers and utilities. Developers are looking for ways to work with utilities rather than against them.

Forward view

Fuel cells are making the difficult transition from concept to reality. Future funding is dependent on solid results, not just promises, so the developers are working to demonstrate their status. Many manufacturers are selling systems for field testing.

The future looks bright for fuel cell technologies. As natural gas becomes the fuel of choice for more and more energy consumers, and with electricity prices becoming more volatile, and as power quality gains ground as a critical issue in the digital society, end users are becoming more educated about the benefits of combined heat and power.

Environmental sensitivity is increasing, further helping to raise the popularity of this clean technology.

Economics and reliability will be deciding factors in culling the winners from the losers in this technology race.

Case Study: Wabash River and Clean Coal Fuel Cell Trials

FCE is proud of its 2 MW fuel cell power plant operating on coal-derived gas as part of the Clean Coal V program from the U.S. DOE. The power plant is based on the direct fuel cell (DFC) technology developed by FCE. It is installed at the Global Energy Wabash River Energy gasification facility near Terre Haute, Indiana (Fig. 7–4).

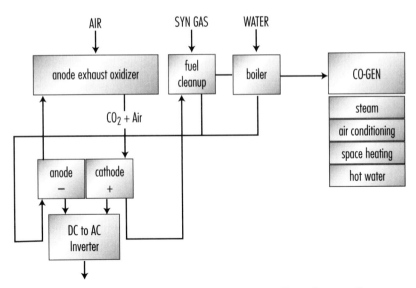

FIG. 7–4 SCHEMATIC OF CARBONATE FUEL CELL POWER PLANT

The carbonate fuel cell uses potassium and lithium carbonates as electrolyte. Syngas from the gasification plant clean-up system is cleaned further and moisturized. The syngas is then fed to the anode side of the fuel cell where methane is internally reformed and CO is shifted to CO_2 and H_2. Spent fuel exits the anode and is consumed in the anode exhaust oxidizer to supply oxygen and CO_2 to the cathode. The resulting electrochemical reactions in the fuel cell anode and cathode produce DC output, which is inverted to AC. The cathode exhaust supplies heat to the fuel cleanup, steam boiler, and cogen systems as it is vented from the plant.

The fuel cell power plant in the Wabash demonstration uses two fuel cell modules, each housing four fuel cell stacks producing DC power. An inverter converts the DC to AC power. The balance of plant equipment includes fuel processing, thermal management, water treatment, instrument air system, and controls. Additional syngas processing equipment is needed to ensure the syngas is suitable for fuel cell operation.

Fuel cell systems operating on coal have been studied extensively over the past few years. Gasification is used to convert the solid fuel to a gas, which is processed to remove sulfur compounds, tars, particulates, and trace contaminants. The cleaned fuel gas is converted to electricity in the fuel cell. Waste heat from the carbonate fuel cell is used to generate steam needed for the gasification process and to generation additional power in a bottoming cycle.

At a 200 MW scale, past studies indicated that using conventional gasification and clean-up technologies, a heat rate of 7186 Btu/kWh, or 47.5% higher heating value (HHV) efficiency, can be achieved with integrated gasification fuel cell (IGFC) plants. Higher efficiencies, 51.7% to 53.5%, could be achieved with higher methane producing gasifiers and by using hot gas cleanup.

The IGFC technology has extremely low emissions, well below any current or anticipated future standards. CO_2 emissions run 1.54 lb/kWh.

Case Study: European Residential Promise

Several companies, including Vaillant in Germany, Sulzer Hexis in Switzerland, PlugPower of the United States, and the Fuel Cell Initiative, a consortium of EWE AG, MVV AG, Ruhrgas AG, and Verbundnetz Gas AG, are working to develop fuel cells for the residential market (Fig. 7–5).

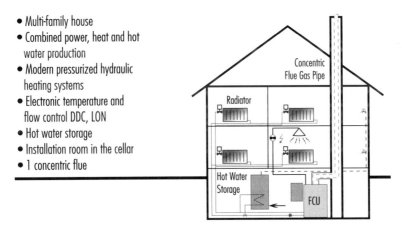

FIG. 7–5 VAILLANT'S EUROPEAN FUEL CELL HEATING APPLIANCE (FCHA) DESIGN

Fuel cells offer a variety of benefits to consumers. They are a cleaner and more reliable power source than conventional power plants. Fuel cells can also offer homeowners an alternative to joining the main grid, which is a bonus for homes in remote locations.

Most of the companies are working on natural gas–powered fuel cells since H_2 is still not a viable energy commodity. The natural gas distribution infrastructure is already in place throughout most industrialized nations. Many homes are already connected to natural gas, which would make the conversion to fuel cell power much easier.

The two most popular fuel cell technologies for residential applications are the PEM and solid oxide.

Sulzer Hexis has designed a SOFC to meet the heat and electricity needs of a single-family residence. The excess heat from the fuel cell is collected and coupled with an auxiliary burner and fed to a water storage unit. This combined conversion allows for efficient fuel use.

The cell stack consists of several repeat elements connected in series. These generate 1 kW of electricity. The auxiliary burner is switched on automatically when needed. The Sulzer Hexis cells have undergone six field tests in Switzerland, Germany, Japan, the Netherlands, and Spain, with more than 75,000 operating hours. Sulzer Hexis is trying to form distribution partnerships with utilities, which will own and operate the systems during the market entry phase.

Plug Power is focusing on its PEM system. Plug Power entered the residential market in 1998 when it demonstrated the Plug Power 7000, a 7 kW residential power system at a home in upstate New York. Since then, the system has run on a regular basis, and in 1999, the home was converted to natural gas operation. Plug Power is also working with GE Power Systems setting up a joint venture with Vaillant. Plug and Vaillant are field testing their FCHA, which will also produce electricity as a byproduct. They are looking to distribute the fuel cell product in multi-family homes across Europe where natural gas is widely available and where environmental protection is a big concern.

Vaillant began developing its PEM fuel cell system in 1999 and hopes to start commercial sales across Europe by 2004. The Vaillant devices will generate 4.5 kW of electrical power and 35 kW of heat, enough to meet the power and heat needs of a multi-family building. Vaillant plans to develop smaller units for single-family homes in the future.

By 2010, Vaillant is predicting 250,000 such units will be installed in Europe, with Vaillant holding a 40% market share.

The actual numbers remain to be seen, but one thing is certain—environmental concerns, reliability, and pricing issues are narrowing the commercialization gap between the sliding capital costs of fuel cells and the convenience of utility heat and power.

H_2 issues

Safe storage of H_2 is an ongoing concern. Storage in carbon and metal hydride or hydride slurries, as well as more conventional pressurized gas storage, is being researched and developed.

Hydrogen earned a bad reputation as an unsafe fuel in part because of the Hindenburg incident. Certainly, H_2 played an important part in the tragedy, but some believe now the airship would have burned anyway. Retired NASA safety expert Addison Bain points to the construction of the ship's gas bags. The bags were made of cellulose acetate or cellulose nitrate, both flammable. Additionally, the bags were coated with aluminum flakes to reflect the sunlight for thermodynamic and aesthetic purposes. Cellulose nitrate and metal chips are ingredients in rocket fuel. Perhaps the dirigible would have burned even had it been filled with inert helium.

These observations are not intended to display a cavalier attitude about the potential hazard inherent in H_2 gas. Surely we all recognize the dangers of gasoline in both liquid and vapor form. All flammables and combustibles must be treated with a certain amount of respect. The fact that H_2 gas dissipates so quickly may actually decrease post-accident fire hazard.

Scientists are experimenting with different designs for H_2 storage, making storage safer and more economical.

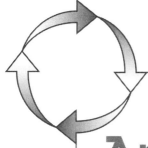

Appendix A

Industry Contact List

Domestic

Allied Utility Network
1800 Peachtree St., NW, Ste. 401
Atlanta, GA 30367
770-972-0611

Alternative Fuels Data Center
P.O. Box 12316
Arlington, VA 22209
800-423-1363

American Bioenergy Association
314 Massachusetts Ave., NE
Washington, D.C. 20002
Phone and Fax: 703-516-4444

American National Standards Institute
11 W. 42nd St.
New York, NY 10036
212-642-4900

American Society for Quality Control
Energy & Environmental Division
P.O. Box 3005
611 E. Wisconsin Ave.
Milwaukee, WI 53202
414-272-8575

American Solar Energy Society
2400 Central Ave., Ste. G-1
Boulder, CO 80301
303-443-3130

American Wind Energy Association
122 C St., NW, 4th Floor
Washington, D.C. 20001
202-383-2500

ASME International
345 E. 47th St.
New York, NY 10017
212-705-7722

Association of Energy Engineers
4025 Pleasantdale Rd., Ste. 420
Atlanta, GA 30340
770-447-5083

Biomass Energy Research Association
116 E St., SE
Washington, D.C. 20003
800-247-1755

Biomass Research & Development Initiative
U.S. Department of Energy
1000 Independence Ave., SW
Washington, D.C. 20585

Biotechnology Center for Fuels and Chemicals
National Renewable Energy Laboratory
1617 Cole Blvd.
Golden, CO 80401

Center for Renewable Energy and Sustainable Technology
1200 Eighteenth St. NW, Ste. 900
Washington, D.C. 20036
202-530-2230

Council on Environmental Quality
722 Jackson Pl., NW
Washington, D.C. 20503
202-395-5750

Department of Energy, U.S.
1000 Independent Ave., SW
Washington, D.C. 20585
202-622-2000

DOE's Energy Efficiency and Renewable Energy Clearinghouse (EREC)
P.O. Box 3048
Merrifield, VA 22116
1-800-DOE-EREC (363-3732)

Appendix A

Electric Generation Association
1401 H St., NW, Ste. 760
Washington, D.C. 20005
202-789-7200

Electric Light & Power **Magazine**
1421 S. Sheridan Rd.
Tulsa, OK 74112
918-835-8161

Electric Power Supply Association
1401 H St., NW, Ste. 760
Washington, D.C. 20005
202-789-7200

Electrical Generating Association
2101 L St., NW, Ste. 405
Washington, D.C. 20037
202-965-1134

Electrical Generating Systems Association
10251 W. Sample Rd., Ste. B
Coral Springs, FL 33065
954-755-AMPS

Electrical Generating Systems Association-EGSA
1650 S. Dixie Highway, 5th Floor
Boca Raton, FL 33432
561-338-3495

Energy Federation Inc.
14 Tech Circle
Natick, MA 01760
508-653-4299

Energy Information Administration
Forrestal Building, Room 1F-048
Washington, D.C. 20585
202-586-8800

Energy Research Institute
6850 Rattlesnake Hammock Rd.
Naples, FL 34113
941-793-1260

Environmental Protection Agency
401 M St. SW
Washington, D.C. 20460
202-260-2090

EPRI
3412 Hillview Ave.
Palo Alto, CA 94304
800-313-3774

Farm Service Agency's Bioenergy Program
U.S. Department of Agriculture
1400 Independence Ave., SW
Washington, D.C. 20250-0506
202-690-0474

Federal Energy Regulatory Commission
888 First St., NE
Washington, D.C. 20426
202-208-0200

Geothermal Resources Council
2001 Second St., Box 1350
Davis, CA 95617
916-758-2360

IEEE Power Engineering Society
P.O. Box 1331
445 Hoes Ln.
Piscataway, NJ 08855
908-981-0060

**Institute of Electrical and
Electronics Engineers Inc.**
345 E. 47th St.
New York, NY 10017
212-705-7900

**International Association for
 Energy Economics**
28790 Chagrin Blvd., Ste. 350
Cleveland, OH 44122
216-464-5365

**International Electrical
Testing Association**
PO Box 687
106 Stone St.
Morrison, CO 80465
303-697-8441

**International League of
Electrical Associations**
2901 Metro Dr., Ste. 203
Bloomington, MN 55425
612-854-4405

**National Association of
Energy Service Companies**
1615 M St., NW, #800
Washington, D.C. 20036
202-822-0950

**National Association of Power
Engineers Inc.**
One Springfield St.
Chicopee, MA 01013
413-592-6273

**National Association of Regulatory
Utility Commissioners**
P.O. Box 684
1100 Pennsylvania Ave., NW, Ste. 603
Washington, D.C. 20044
202-898-2200

**National Bioenergy
Industries Association**
122 C St., NW, 4th Floor
Washington, D.C. 20001
202-383-2540

**National Electrical
Manufacturers Association**
1300 N. 17th St., Ste. 1847
Rosslyn, VA 22209
703-841-3200

National Energy Information Center
1000 Independence Ave., SW
Washington, D.C. 20585
202-586-8800

**National Independent
Energy Producers**
601 Thirteenth St., NW, Ste. 320
Washington, D.C. 20005
202-793-6506

Appendix A

National Hydropower Association
122 C St., NW, 4th Floor
Washington, D.C. 20001
202-383-2530

**National Renewable
Energy Laboratory**
1617 Cole Blvd.
Golden, CO 80401
303-275-3000

National Science Foundation
4201 Wilson Blvd.
Arlington, VA 22203
703-306-1224

**North American Electric
Reliability Council**
Princeton Forrestal Village
116–390 Village Blvd.
Princeton, NJ 08540
609-452-8060

Oak Ridge National Laboratory
P.O. Box 2008, MS-6422
Oak Ridge, TN 37831-6422

**Occupational Safety and
Health Administration**
200 Constitution Ave., NW, Room S2315
Washington, D.C. 20210
202-219-6091

**Office of Energy Efficiency and
Renewable Energy**
U.S. Department of Energy
www.eren.doe.gov

**Power Electronics
Applications Center**
10521 Research Dr., Ste. 400
Knoxville, TN 37932
423-974-8288

Power Engineering **Magazine**
1421 S. Sheridan Rd.
Tulsa, OK 74112
918-835-8161

Power Marketing Association
1519 Twenty-second St., Ste. 200
Arlington, VA 22202
703-892-0010

Renewable Fuels Association
One Massachusetts Ave., Ste. 820
Washington, D.C. 20001
202-289-3835

Solar Energy Industries Association
122 C St., NW, 4th Floor
Washington, D.C. 20001
202-383-2600

**United Bioenergy
Commercialization Association**
7164 Gateway Dr.
Columbia, MD. 21044
410-953-6252

United States Energy Association
1620 Eye St., NW, Ste. 1000
Washington, D.C. 20006
202-331-0415

Utility Marketing Association
4755 Walnut St.
Boulder, CO 80301
303-786-7444

International

Canadian Electrical Association
1155 Metcalfe St.
Sun Life Building
Montreal, Quebec, Canada H3A 2V6
514-866-6121

Canadian Electricity Association
1 Westmount Square, Ste. 1600
Montreal, Canada H3Z 2P9
514-937-6181

Canadian Energy Research Institute
3512–33 St., NW, Ste. 150
Calgary, Canada T2L 2A6
403-282-1231

Canadian Standards Association
178 Rexdale Blvd.
Etobicoke, Canada M9W IR3
416-747-4000
800-473-6726

Electricity Association
30 Millbank
London, U.K. SWIP 4RD
+44 (0) 171-963-5700

Electricity Supply Association of Australia Ltd.
Level 11
74 Castlereagh St.
Sydney, Australia NSW 2000
+61 2-9233-7222

Finnish Energy Industries Federation, Finergy
P.O. Box 21
Etelaranta 10
Helsinki, Finland
+358 9-686-161

Industrial and Power Association
Brunel Building
Scottish Enterprise Technology Park
East Kilbride, Scotland, U.K. G75 0QD
+44 (0) 1355-272-630

Institution of Electrical Engineers
Michael Faraday House
Stevenage, Herts, U.K. SGI 2AY
+44 (0) 1438-767-249

International Energy Agency
9 Rue de la Federation
75739 Paris Cedex 15
France
331-4057-6551

Latin American Association of Development Financing Institutions
Paseo de la Republica
3211, Lima 27, Peru
511-442-2400

Norwegian Electric Power Research Institute
Sem Saelandsvei 11, N-7465
Trondheim, Norway
+47 73-59-7200

Secretariat for Central American Economic Integration
4a Avenida 10–25, Zona 14
Ciudad de Guatemala, Guatemala
+502 368-2151-54

South African Institute of Electrical Engineering
P.O. Box 93541
Yeoville, 2143
Johannesburg, South Africa
+27 11-487-3003

Spanish Electricity Industry Association
Francisco Gervas No. 3
28020 Madrid, Spain
+1 941-793-1922

World Energy Efficiency Association
1400 16th St. NW, Ste. 340
Washington, D.C. 20036, USA
202-797-6570

World Energy Council
5th Floor, Regency House 1
4 Warwick St.
London, U.K. WIR 6LE+44 20-7734-5996

World Trade Organization
Centre William Rappard
Rue de Lausanne 154
CH-1211 Geneva 21
Switzerland
+41 22-739-5111

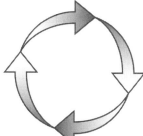

Appendix B

Renewable Energy Glossary

A

absorber. The component of a solar thermal collector that absorbs solar radiation and converts it to heat, or as in a solar PV device, the material that readily absorbs photons to generate charge carriers (free electrons or holes).

absorption. The passing of a substance or force into the body of another substance.

absorption chiller. A type of air-cooling device that uses absorption cooling to cool interior spaces.

absorption cooling. A process in which cooling of an interior space is accomplished by the evaporation of a volatile fluid, which is then absorbed in a strong solution, desorbed under pressure by a heat source, and then recondensed at a temperature high enough that the heat of condensation can be rejected to a exterior space.

accumulator. A component of a heat pump storing liquid and keeping it from flooding the compressor. The accumulator takes the strain off the compressor and improves the reliability of the system.

active solar heater. A solar water or space-heating system using pumps or fans to circulate the fluid (water or heat-transfer fluid like diluted antifreeze) from the solar collectors to a storage tank subsystem.

aerobic bacteria. Microorganisms that require free oxygen (or air) to live, and contribute to the decomposition of organic material in soil or composting systems.

affiliated power producer. A company that generating power and is affiliated with a utility.

air collector. In solar heating systems, a type of solar collector in which air is heated in the collector.

air monitoring. Intermittent or continuous testing of emission air for pollution levels.

air pollution. Contaminants in the atmosphere that have toxic characteristics and which are believed to be harmful to the health of animal or plant life.

air pollution abatement equipment. Equipment used to reduce or eliminate airborne pollutants, including particulate matter such as dust, smoke, fly ash, or dirt and sulfur oxides, NO_x, CO, hydrocarbons, odors, and other pollutants. Examples of air pollution abatement structures and equipment include flue-gas particulate collectors and NO_x control devices.

air quality. Air quality is determined by the amount of pollutants and contaminants present.

air quality standards. Pollutant limitations defined by law. Standards vary by country and are generally established by federal governments.

albedo. The ratio of light reflected by a surface to the light falling on it.

alcohol. A group of organic compounds composed of carbon, hydrogen (H_2), and oxygen; a series of molecules composed of a hydrocarbon plus a hydroxyl group; includes methanol, ethanol, isopropyl alcohol, and others.

alkaline fuel cell. Fuel cells that use alkaline potassium hydroxide as the electrolyte. These are currently far too expensive for commercial use, but NASA uses them.

allowable emissions. The emissions rate of a stationary source calculated using the maximum rated capacity of the source and the most stringent applicable governmental standards.

allowance. An authorization to emit, during or after a specified calendar year, one ton of sulfur dioxide.

allowance trading. Buying or selling emission permits for sulfur dioxide. The Clean Air Act Amendments of 1990 require most fossil-fuel electric generating facilities to have allowances for each ton of sulfur dioxide emission produced. Utilities are allowed to buy and sell these allowances, which the EPA issues annually based on each facility's historic fuel usage and other factors.

alternating current (AC). A periodic current, the average value of which over a period is zero. Unless distinctly specified otherwise, the term refers to a current that reverses its direction at regularly recurring intervals of time and has alternately positive and negative values. Almost all electric utilities generate AC electricity because it can easily be transformed to higher or lower voltages.

alternative fuel. According to the stipulations of the Energy Information Administration, alternative fuels can include methanol; denatured ethanol and other alcohols; fuel mixtures containing 85% or more by volume of methanol, denatured ethanol, and other alcohols with gasolines or other fuels; liquefied petroleum gas (propane); H_2; coal derived liquid fuels; fuels other than alcohol derived from biological materials including biofuels such as soy-diesel; and electricity including electricity from solar energy. Also any other fuel that is substantially not petroleum and would yield substantial energy security benefits and substantial environmental benefits. The term alternative fuel does not include alcohol or other blended portions of primarily petroleum-based fuels used as oxygenates or extenders or the 10% ethanol portion of gasohol.

ambient conditions. The outside weather conditions, including temperature, humidity, and barometric pressure. Ambient conditions can affect the available capacity of some types of power plants.

American Public Power Association (APPA). The trade association of publicly held power entities.

American Society for Testing and Materials (ASTM). The entity responsible for the issue of many standard methods used in the energy industry.

American Society of Heating, Refrigeration, and Air-Conditioning Engineers (ASHRAE).

ampere. A unit of measurement of electric current produced in a circuit by one volt acting through a resistance of one ohm.

anaerobic bacteria. Microorganisms that live in oxygen deprived environments.

anaerobic digestion. The complex process by which organic matter is decomposed by anaerobic bacteria. The decomposition process produces a gaseous byproduct often called *biogas* primarily composed of methane, carbon dioxide (CO_2), and hydrogen sulfide.

anaerobic digester. A device for optimizing the anaerobic digestion of biomass and/or animal manure, and possibly to recover biogas for energy production. Digester types include batch, complete mix, continuous flow (horizontal or plug-flow, multiple-tank, and vertical tank), and covered lagoon.

anhydrous ethanol. One hundred percent alcohol; neat ethanol.

anemometer. An instrument for measuring the force or velocity of wind; a wind gauge.

angle of incidence. In reference to solar energy systems, the angle at which direct sunlight strikes a surface; the angle between the direction of the sun and perpendicular to the surface. Sunlight with an incident angle of 90 degrees tends to be absorbed, while lower angles tend to be reflected.

angle of inclination. In reference to solar energy systems, the angle that a solar collector is positioned above horizontal.

annual demand. The greatest electric power demand occurring in any calendar year.

annual operating time. Also called annual service hours. The number of hours per year during which a unit or group of units is operated. Can be continuous or interrupted operations.

antireflection coating. A thin coating of a material applied to a PV cell surface that reduces the light reflection and increases light transmission.

aperture. An opening; in solar collectors, the area through which solar radiation is admitted and directed to the absorber.

array (solar). Any number of solar PV modules, solar thermal collectors, or reflectors connected together to provide electrical or thermal energy.

attainment area. A geographic area under the Clean Air Act that is in compliance with the Act's national Ambient Air Quality Standards. This designation is made on a pollution-specific basis.

availability. The unit of measure for the actual time a transmission line or generating unit is capable of providing service, if needed.

average revenue per kwh. The average revenue per kwh of electricity sold by sector (residential, commercial, industrial, or other) and geographic area (state, census division, and national) is calculated by dividing the total monthly revenue by the corresponding total monthly sales for each sector and geographic area.

avoided cost. A utility company's production or transmission cost avoided by conservation or purchasing from another source rather than by building a new generation facility.

B

backup power. Power supplied to a customer when its normal supply is interrupted.

bagasse. The fibrous material remaining after the extraction of juice from sugarcane; often burned by sugar mills as a source of energy.

barrel. A volumetric unit of measure for crude oil and petroleum products equivalent to 42 U.S. gallons.

baseload. The minimum amount of electric power or natural gas delivered or required over a given period of time at a steady rate. The lowest load level during a utility's daily or annual cycle.

baseload capacity. The generating equipment normally operated to serve loads on an around-the-clock basis.

baseload plant. A plant, usually housing high-efficiency steam-electric units, which is normally operated to take all or part of the minimum load of a system, and which consequently produces electricity at an essentially constant rate and runs continuously. These units are operated to maximize system mechanical and thermal efficiency and minimize system operating costs.

bcf. Billion cubic feet.

binary cycle. Combination of two power plant turbine cycles utilizing two different working fluids for power production. The waste heat from the first turbine cycle provides the heat energy for the operation of the second turbine, thus providing higher overall system efficiencies.

binary cycle geothermal plants. Binary cycle systems can be used with liquids at temperatures less than 350°F (177°C). In these systems, the hot geothermal liquid vaporizes a secondary working fluid, which then drives a turbine.

bioconversion. The conversion of one form of energy into another by the action of plants or microorganisms. The conversion of biomass to ethanol, methanol, or methane.

bioenergy. The conversion of the complex carbohydrates in organic material into energy.

biogas. A combustible gas created by anaerobic decomposition of organic material, composed primarily of methane, CO_2, and hydrogen sulfide.

biogasification or biomethanization. The process of decomposing biomass with anaerobic bacteria to produce biogas.

biomass. As defined by the Energy Security Act (PL 96-294) of 1980, *any organic matter that is available on a renewable basis, including agricultural crops and agricultural wastes and residues, wood and wood wastes and residues, animal wastes, municipal wastes, and aquatic plants.*

biomass energy. Energy produced by the conversion of biomass directly to heat or to a liquid or gas that can be converted to energy.

biomass fuel. Biomass converted directly to energy or converted to liquid or gaseous fuels such as ethanol, methanol, methane, and H_2.

biomass gasification. The conversion of biomass into a gas, by biogasification (see above) or thermal gasification, in which H_2 is produced from high-temperature gasifying and low-temperature pyrolysis of biomass.

breaker. A device to break the current of a given electric circuit by opening the circuit.

Btu. British thermal unit. A standard unit for measuring the quantity of heat energy equal to the quantity of heat required to raise the temperature of a pound of water by one degree Fahrenheit.

C

crystalline silicon PV cell. A type of PV cell made from a single crystal or a polycrystalline slice of silicon. Crystalline silicon cells can be joined together to form a module (or panel).

cut-in-speed. The lowest wind speed at which a wind turbine begins producing usable power.

cutout-speed. The highest wind speed at which a wind turbine stops producing power.

capability. The maximum load that a generating unit, generating station, or other electrical apparatus can carry under specified conditions for a given period of time without exceeding approved limits of temperature and stress.

capacitor bank. An assembly of capacitors and all necessary accessories, such as switching equipment, protective equipment, controls, and other devices needed for a complete operating installation.

capacity charge. An element in a two-part pricing method used in capacity transactions (energy charge is the other element). The capacity charge, sometimes called demand charge, is assessed on the amount of capacity being purchased.

capacity. The amount of electric power delivered or required for which a generator turbine, transformer, transmission circuit, station, or system is rated by the manufacturer.

carbon dioxide (CO_2). A colorless, odorless, nonpoisonous gas that occurs in ambient air. It is produced by fossil fuel combustion or the decay of materials.

carbon monoxide (CO). A colorless, odorless, tasteless, but poisonous gas produced mainly from the incomplete combustion of fossil fuels.

cathode. The negative pole or electrode of an electrolytic cell, vacuum tube, etc., where electrons enter (current leaves) the system; the opposite of an anode.

cellulose. The fundamental constituent of all vegetative tissue; the most abundant material in the world.

central receiver solar power plants. Also known as *power towers*, these use fields of two-axis tracking mirrors known as heliostats. Each heliostat is individually positioned by a computer control system to reflect the sun's rays to a tower-mounted thermal receiver. The effect of many heliostats reflecting to a common point creates the combined energy of thousands of suns, which produces high-temperature thermal energy. In the receiver, molten nitrate salts absorb the heat energy. The hot salt is then used to boil water to steam, which is sent to a conventional steam turbine-generator to produce electricity.

chemical vapor deposition (CVD). A method of depositing thin semi-conductor films used to make certain types of solar PV devices. With this method, a substrate is exposed to one or more vaporized compounds, one or more of which contain desirable constituents. A chemical reaction is initiated, at or near the substrate surface, to produce the desired material that will condense on the substrate.

chlorofluorocarbons. A family of inert, nontoxic, easily liquefied chemicals used in refrigeration, air conditioning, packaging, and insulation, or as solvents or aerosol propellants. They are thought to be major contributors to potential ozone thinning and global warming and are therefore becoming increasingly regulated and their use more restricted.

circuit. A circuit is a conductor or system of conductors forming a closed path through which electric current flows.

circulating fluidized bed. A type of furnace or reactor in which the emission of sulfur compounds is lowered by the addition of crushed limestone in the fluidized bed thus obviating the need for much of the expensive stack gas clean-up equipment. The particles are collected and recirculated, after passing through a conventional bed, and cooled by boiler internals.

closed-loop geothermal heat pump systems. Closed-loop (also known as *indirect*) systems circulate a solution of water and antifreeze through a series of sealed loops of piping. Once the heat has been transferred into or out of the solution, the solution is recirculated. The loops can be installed in the ground horizontally or vertically, or they can be placed in a body of water, such as a pond.

closed-loop biomass. Any organic matter from a plant that is planted for the exclusive purpose of being used to produce energy. This does not include wood or agricultural wastes or standing timber.

cogeneration. The simultaneous production of power and thermal energy, such as burning natural gas to produce electricity and using the heat produced to create steam for industrial use.

combined cycle. An electric generating technology in which additional electricity is produced from otherwise lost waste heat exiting from the gas turbines. The exiting heat is routed to a conventional boiler or to a heat recovery steam generator for utilization by a steam turbine in the production of electricity. The process increases the efficiency of the electricity generating unit.

commercial customers. A statistical and regulatory category of energy use, embracing retail and wholesale trade, service establishments, hotels, offices, public institutions, and sometimes apartments that are separately metered.

commercial financing. Money obtained from banks and other commercial institutions.

compressed air storage. The storage of compressed air in a container for use to operate a prime mover for electricity generation.

concentrating (solar) collector. A solar collector that uses reflective surfaces to concentrate sunlight onto a small area, where it is absorbed and converted to heat or, in the case of solar PV devices, into electricity. Concentrators can increase the power flux of sunlight hundreds of times. The principal types of concentrating collectors include: compound parabolic, parabolic trough, fixed reflector moving receiver, fixed receiver moving reflector, Fresnel lens, and central receiver. A PV concentrating module uses optical elements (Fresnel lens) to increase the amount of sunlight incident onto a PV cell. Concentrating PV modules/arrays must track the sun and use only the direct sunlight because the diffuse portion cannot be focused onto the PV cells. Concentrating collectors for home or small business solar water heating applications are usually parabolic troughs that concentrate the sun's energy on an absorber tube (called a receiver), which contains a heat-transfer fluid.

condenser. Condensers are equipment in generating facilities that capture steam and turn it back into water for reuse in the feedwater system of the plant.

constant-speed wind turbines. Wind turbines that operate at a constant rotor revolutions per minute (rpm) and are optimized for energy capture at a given rotor diameter at a particular speed in the wind power curve.

convection. The transfer of heat by means of air currents.

convergence. The coming together and merging of previously distinct industries. This phenomenon is currently under way for the electricity and fuels industries, particularly electricity and natural gas.

cooling tower. The portion of a power facility's water circulating system that extracts the heat from water coming out of the plant's condenser, cooling it down and transferring the heat into the air while the water returns through the system to become boiler make-up water.

cooperative electric utility. An electric utility legally established to be owned by and operated for the benefit of those using its service. The utility company will generate, transmit, and/or distribute supplies of electric energy to a specified area not being serviced by another utility. Such ventures are generally exempt from federal income tax laws. Most electric cooperatives have been initially financed by the Rural Electrification Administration, U.S. Department of Agriculture.

credit rating. A classification of the level of credit ascribed to a person, company, or country.

creditability. The ability of a person, company, or country to obtain credit.

credit worthiness. Similar to creditability and credit rating, a company's creditworthiness is a judgment of its trustworthiness and its ability to repay a loan.

current. The flow of electrons in an electrical conductor. The rate of movement of the electricity, measured in amperes.

customer density. Number of customers in a given unit of area or on a given length of distribution line.

D

Darrius (wind) machine. A type of vertical-axis wind machine that has long, thin blades in the shape of loops connected to the top and bottom of the axle; often called an *eggbeater windmill*.

declination. The angular position of the sun at solar noon with respect to the plane of the equator.

decomposition. The process of breaking down organic material; reduction of the net energy level and change in physical and chemical composition of organic material.

degree day. A unit for measuring the extent that the outdoor daily average temperature (the mean of the maximum and minimum daily dry-bulb temperatures) falls below (in the case of heating, see heating degree day), or falls above (in the case of cooling, see cooling degree day) an assumed base temperature, normally taken as 65°F, unless otherwise stated. One degree day is counted for each degree below (for heating) or above (in the case of cooling) the base, for each calendar day on which the temperature goes below or above the base.

demand. Ability and willingness of customers to purchase a product or service. In electricity, the rate at which electric energy is delivered to or by a system, part of a system, or piece of equipment, at a given instant or averaged over any designated period of time.

demand-side management (DSM). The term for all activities or programs undertaken by an electric system or its customers to influence the amount and timing of electricity use. Included in DSM are the planning, implementation, and monitoring of utility activities that are designed to influence consumers use of electricity in ways that will produce desired changes in a utility's load shape. These programs are dwindling, and expected to experience a great decline under deregulation.

Department of Energy (DOE). Established in 1977, the DOE manages programs of research, development and commercialization for various energy technologies, and associated environmental, regulatory, and defense programs. DOE promulgates energy policies and acts as a principal adviser to the president on energy matters.

deregulation. Relaxing or eliminating laws and regulations controlling an industry or industries.

diffuse solar radiation. Sunlight scattered by atmospheric particles and gases so it arrives at the earth's surface from all directions and cannot be focused.

digester (anaerobic). A device in which organic material is biochemically decomposed (digested) by anaerobic bacteria to treat the material and/or to produce biogas.

direct beam radiation. Solar radiation that arrives in a straight line from the sun.

direct current (DC). An electric current that flows in one direction with a magnitude that does not vary or that varies only slightly.

distribution automation. A system consisting of line equipment, communications infrastructure, and information technology that is used to gather intelligence about the distribution system and provide analysis and control in order to maximize operating efficiency and reliability. It includes small distribution substations, sub-transmission and distribution feeder reclosers, regulators, and sectionalizers, which can be remotely monitored and controlled.

distribution system. For electricity, the substations, transformers, and lines conveying electricity from the generation site to the consumer.

downwind wind turbine. A horizontal axis wind turbine in which the rotor is downwind of the tower.

dry steam geothermal plants. Conventional turbine generators are used with the dry steam resources. The steam is used directly, eliminating the need for boilers and boiler fuel that characterizes other steam-power-generating technologies. This technology is limited because dry-steam hydrothermal resources are extremely rare. The Geysers, in California, is the nation's only dry steam field.

E

Edison Electric Institute (EEI). The association of the investor-owned electric utilities in the United States and industry affiliates worldwide. Its U.S. members serve almost all of the customers served by the investor-owned segment of the electric utility industry. They generate almost 80%

of all electricity generated by utilities and service more than 75% of all customers in the nation. EEI's basic objective is the *advancement in the public service of the art of producing, transmitting, and distributing electricity and the promotion of scientific research in such field.* EEI compiles data and statistics relating to the industry and makes them available to member companies, the public, and government representatives.

electric and magnetic fields (EMF). Electric and magnetic fields are created when energy flows through an energized conductor. The electric field is from the voltage impressed on the conductors and the magnetic field is from the current in the conductors. These fields surround the conductors. Electric fields are measured in volts per meter or kilovolts per meter and magnetic fields are measured in gauss or tesla. Electric and magnetic fields occur naturally, but can also be created. There is debate regarding possible health effects of these fields when they occur in proximity to residences.

electric capacity. The ability of a power plant to produce a given output of electric energy at an instant in time. Capacity is measured in kw or MW.

electric current. A flow of electrons in an electrical conductor. The strength or rate of movement of the electricity is measured in amperes.

electric rate schedule. A statement of the electric rate and the terms and conditions governing its application, including attendant contract terms and conditions that have been accepted by a regulatory body with appropriate oversight authority.

electric utility. A corporation, person, agency, authority, or other legal entity or instrumentality that owns and/or operates facilities within the United States, its territories, or Puerto Rico for the generation, transmission, distribution, or sale of electric energy primarily for use by the public and files forms listed in the Code of Federal Regulations, Title 18, Part 141. Facilities that qualify as cogenerators or small power producers under the PURPA are not considered electric utilities.

electricity. The flow of electrons in a conducting material. The flow is called a current.

electrolysis. A chemical change in a substance resulting from the passage of an electric current through an electrolyte. The production of commercial H_2 by separating the elements of water, H_2, and oxygen, by charging the water with an electrical current.

electrolyte. A nonmetallic (liquid or solid) conductor that carries current by the movement of ions (instead of electrons) with the liberation of matter at the electrodes of an electrochemical cell.

emissions. Any waste products leaving a power plant. This term generally applies to air pollution, but it can also apply to soil or water waste issues. There are many substances that can be emitted from power plants, and most of them are regulated and monitored.

end-user. The ultimate consumer, as opposed to a customer purchasing for resale.

energy charge. The portion of the charge for electric services that is based on the electric energy either consumed or billed.

energy crops. Crops grown specifically for their fuel value. These include food crops such as corn and sugar cane, and nonfood crops such as poplar trees and switchgrass. Currently, two energy crops are under development: short-rotation woody crops, which are fast-growing hardwood trees harvested in five to eight years; and herbaceous energy crops, such as perennial grasses, which are harvested annually after taking two to three years to reach full productivity.

energy efficiency. Refers to programs that are aimed at reducing the energy used by specific end-use devices and systems, typically without affecting the services provided. These programs reduce overall electricity consumption (reported in megawatt-hours), often without explicit consideration for the timing of program-induced savings. Such savings are generally achieved by substituting technically more advanced equipment to produce the same level of end-use services (e.g., lighting, heating, motor drive) with less electricity. Examples include high-efficiency appliances, efficient lighting programs, high-efficiency HVAC systems or control modifications, efficient building design, advanced electric motor drives, and heat recovery systems.

energy. Power is the capability of doing work. Energy is power supplied over time, expressed in kwh. Energy can take on different forms, some of which are easily convertible and can be changed to another form useful for work. Most of the world's convertible energy comes from fossil fuels that are burned to produce heat that is then used as a transfer medium to medium to mechanical or other means in order to accomplish tasks. Electrical energy is usually measured in kwh, while heat energy is generally measured in British thermal units.

energy storage. The process of storing, or converting energy from one form to another, for later use; storage devices and systems include batteries, conventional and pumped storage hydroelectric, flywheels, compressed gas, and thermal mass.

Environmental Protection Agency (EPA). This agency administers federal environmental policies, enforces environmental laws and regulations, performs research, and provides information on environmental subjects.

equity. Ownership right in a property. Equity sponsors own a share of a project, as opposed to debt sponsors, which lend money to the project.

ethanol. Ethyl alcohol (C_2H_5OH) is a colorless liquid that is the product of fermentation used in alcoholic beverages, industrial processes, and as a fuel additive. Also known as grain alcohol.

ethyl tertiary butyl ether (ETBE). A chemical compound produced in a reaction between ethanol and isobutylene (a petroleum-derived by-product of the refining process). ETBE has characteristics superior to other ethers: low volatility, low water solubility, high-octane value, and a large reduction in CO and hydrocarbon emissions.

economic viability. Having a reasonable chance of remaining financially solvent and generating a profit.

F

feather. In a wind energy conversion system, to pitch the turbine blades reducing their lift capacity as a method of shutting down the turbine during high wind speeds.

federal electric utilities. A classification of utilities that applies to those agencies of the federal government involved in the generation and/or transmission of electricity. Most of the electricity generated by federal electric utilities is sold at wholesale prices to local government-owned and cooperatively owned utilities, and to investor-owned utilities. These government agencies are the Army Corps of Engineers and the Bureau of Reclamation, which generate electricity at federally owned hydroelectric projects. The TVA produces and transmits electricity in the Tennessee Valley region.

Federal Energy Regulatory Commission (FERC). A quasi-independent regulatory agency within the Department of Energy having jurisdiction over interstate electricity sales, wholesale electric rates, hydroelectric licensing, natural gas pricing, oil pipeline rates, and gas pipeline certification.

Federal Power Act. Enacted in 1920, and amended in 1935, the Act consists of three parts. The first part incorporated the Federal Water Power Act administered by the former Federal Power Commission, whose activities were confined almost entirely to licensing nonfederal hydroelectric projects. Parts II and III were added with the passage of the Public Utility Act. These parts extended the Act's jurisdiction to include regulating the interstate transmission of electrical energy and rates for its sale as wholesale in interstate commerce. The FERC is now charged with the administration of this law.

Federal Power Commission (FPC). The predecessor agency of the FERC. The FPC was created by an Act of Congress under the Federal Water Power Act on June 10, 1920. It was charged originally with regulating the electric power and natural gas industries. The FPC was abolished on September 20, 1977, when the DOE was created. The functions of the FPC were divided between the DOE and the FERC.

flash-steam geothermal plants. When the temperature of the hydrothermal liquids is more than 350°F (177°C), flash-steam technology is generally employed. In these systems, most of the liquid is flashed to steam. The steam is separated from the remaining liquid and used to drive a turbine generator. While the water is returned to the geothermal reservoir, the economics of most hydrothermal flash plants are improved by using a dual-flash cycle, which separates the steam at two different pressures. The dual-flash cycle produces 20% to 30% more power than a single-flash system at the same fluid flow.

flat plate solar thermal/heating collectors. Large, flat boxes with glass covers and dark-colored metal plates inside that absorb and transfer solar energy to a heat transfer fluid. This is the most common type of collector used in solar hot water systems for homes or small businesses.

flat plate solar PV module. An arrangement of PV cells or material mounted on a rigid flat surface with the cells exposed freely to incoming sunlight.

fluidized bed combustion (FBC). A type of furnace or reactor in which fuel particles are combusted while suspended in a stream of hot gas.

force majeure. A contractual provision by which a party's obligations are waived if a superior force, such as weather, war or an act of God, makes it impossible for those obligations to be met.

fossil fuels. Fuels formed in the ground from the remains of dead plants and animals. It takes millions of years to form fossil fuels. Oil, natural gas, and coal are fossil fuels.

Francis turbine. A type of hydropower turbine containing a runner that has water passages through it formed by curved vanes or blades. As the water passes through the runner and over the curved surfaces, it causes rotation of the runner. The rotational motion is transmitted by a shaft to a generator.

Fresnel lens. An optical device for concentrating light that is made of concentric rings facing at different angles so that light falling on any ring is focused to the same point.

fuel cell. A device capable of converting natural gas, H_2, or other gaseous fuels directly into electricity and heat via an electrochemical process that avoids the energy losses associated with combustion and the spinning or reciprocation of mechanical parts.

fuel grade alcohol. Usually refers to ethanol from 160 to 200 proof.

G

gallium arsenide. A compound used to make certain types of solar PV cells.

gasification. The process in which a solid fuel is converted into a gas; also known as pyrolitic distillation or pyrolysis. Production of a clean fuel gas makes a wide variety of power options available.

gasohol. A registered trademark of an agency of the state of Nebraska, for an automotive fuel containing a blend of 10% ethanol and 90% gasoline.

gas turbine. Consists of an axial-flow air compressor and one or more combustion chambers where liquid or gaseous fuel is burned The hot gases that are produced are passed to the turbine where the gases expand to drive the generator and then are used to run the compressor.

generating unit. Any combination of generators, reactors, boilers, combustion turbines, or other prime movers operated together or physically connected to produce electric power.

generation. The process of producing electric energy by transforming other forms of energy. It also refers to the amount of electric energy produced, generally expressed in kwh or MWh.

generator nameplate capacity. The full-load continuous rating of a generator, prime mover, or other electric power production equipment under specific conditions as designated by the manufacturer. Installed generator nameplate rating is usually indicated on a plate physically attached to the generator.

generator. A machine that converts mechanical energy into electrical energy.

geopressurized brines. These brines are hot (300°F to 400°F) (149°C to 204°C) pressurized waters containing dissolved methane and lying at depths of 10,000 ft (3048 m) to more than 20,000 ft (6096 m) below the earth's surface. The best-known geopressured reservoirs lie along the Texas and Louisiana Gulf Coast. At least three types of energy could be obtained: thermal energy from high-temperature fluids; hydraulic energy from the high pressure; and chemical energy from burning the dissolved methane gas.

geothermal energy. Energy produced by the internal heat of the earth; geothermal heat sources include hydrothermal convective systems; pressurized water reservoirs; hot dry rocks; manual gradients; and magma. Geothermal energy can be used directly for heating or to produce electric power.

geothermal heat pump. A type of heat pump that uses the ground, ground water, or ponds as a heat source and heat sink, rather than outside air. Ground or water temperatures are more constant and are warmer in winter and cooler in summer than air temperatures. Geothermal heat pumps operate more efficiently than *conventional* or *air source* heat pumps.

geothermal plant. An electric power plant in which the prime mover is a steam turbine. The turbine is driven either by steam produced from hot water or by natural steam that derives its energy from heat found in rocks or fluids at various depths beneath the surface of the earth. The energy is extracted by drilling and/or pumping.

gigawatt (GW). A unit of electric power equal to one billion watts or one thousand megawatts.

gigawatt-hour (GWh). One billion watt-hours.

global warming. A hypothesized increase in worldwide atmospheric temperatures, caused by an intensification of the natural greenhouse effect thought to attend the socially accelerated accumulation of CO_2 and other heat-retaining gases.

green marketing. Using an ecological perspective in marketing, packaging, or promoting a product as environmentally benign or beneficial.

green power. A popular term for energy produced from clean, renewable energy resources.

green pricing. A practice engaged in by some regulated utilities where electricity produced from clean, renewable resources is sold at a higher cost than that produced from fossil or nuclear power plants, supposedly because some buyers are willing to pay a premium for clean power.

greenfield plant. Refers to a new electric power generating facility built from the ground up on a site that has not been used for industrial uses previously; essentially a plant that starts with a green field. Plants built on sites that have already been used for another power plant or other industrial use are called brownfield plants.

greenhouse effect. The natural warming of the earth's lower atmosphere, associated with solar energy reflected from the surface and retained by water vapor and irradiative gases, including CO_2 and methane.

greenhouse gases. Those gases, such as CO_2, nitrous oxide, and methane, that are transparent to solar radiation but opaque to longwave radiation. Their action in the atmosphere is similar to that of glass in a greenhouse.

grid. The layout of an electrical distribution system.

gross generation. The total amount of electric energy produced by the generating units at a generating station or stations, measured at the generator terminals.

ground. A conducting connection, whether intentional or accidental, by which an electric circuit or equipment is connected to the earth, to some conducting body of relatively large extent that serves in place of the earth.

ground reflection. Solar radiation reflected from the ground onto a solar collector.

H

head. A unit of pressure for a fluid, commonly used in water pumping and hydro power to express height a pump must lift water, or the distance water falls. Total head accounts for friction head losses, etc.

heat rate. A power plant term for the efficiency of the power plant. Heat rate measures how much of the fuel burned actually turns into electricity. Heat rate is generally represented as a mixture of British and metric units, Btu/kWh.

heliostat. A device that tracks the movement of the sun; used to orient solar concentrating systems.

heliothermal. Any process that uses solar radiation to produce useful heat.

hertz (Hz). The international standard unit of frequency, defined as the frequency of a periodic phenomenon with a period of one second. Electricity is generally either 50 Hz or 60 Hz.

horizontal-axis wind turbines. Turbines in which the axis of the rotor's rotation is parallel to the wind stream and the ground.

hot dry rock. A geothermal energy resource that consists of high temperature rocks more than 300°F (150°C) that may be fractured and have little or no water. To extract the heat, the rock must first be fractured, and then water is injected into the rock and pumped out to extract the heat. In the western United States, as much as 95,000 square miles (246,050 square km) have hot dry rock potential.

hub height. The height above the ground that a horizontal axis wind turbine's hub is located.

hybrid system. A renewable energy system including two different types of technologies that produce the same type of energy; e.g., a wind turbine and a solar PV array combined to meet a power demand.

hydroelectric power plant. A power plant that produces electricity by the force of water falling through a hydro turbine that spins a generator.

hydrocarbon. An organic chemical compound of H$_2$ and carbon in gaseous, liquid, or solid phase. The molecular structure of hydrocarbon compounds varies from the simple, such as methane, to the very heavy and very complex.

hydrothermal fluids. These fluids can be either water or steam trapped in fractured or porous rocks; they are found from several hundred feet to several miles below the earth's surface. The temperatures vary from about 90°F to 680°F (32°C to 360°C) but roughly two-thirds range in temperature from 150°F to 250°F (65.5°C to 121.1°C). The latter are the easiest to access and, therefore, the only forms being used commercially.

I

incident solar radiation. The amount of solar radiation striking a surface per unit of time and area.

industrial sector. Electric utilities generally divide customers into classes, broadly, residential, commercial and industrial. The industrial sector includes manufacturing, construction, mining, agriculture and others. Industrial users generally have heavier electrical use than residential or commercial users.

infrared radiation. Electromagnetic radiation whose wavelengths lie in the range from 0.75 micrometer to 1000 micrometers; invisible long wavelength radiation (heat) capable of producing a thermal or PV effect, though less effective than visible light.

insolation. The solar power density incident on a surface of stated area and orientation, usually expressed as watts per square meter or Btu per square foot per hour.

internal combustion plant. A plant in which the prime mover is an internal combustion engine. This type of engine has one or more cylinders, in which the process of combustion takes place, converting energy released from the rapid burning of a fuel-air mixture into mechanical energy. Diesel or gasoline engines are the principal types used in electric plants. These plants are generally used during periods of high electricity demand as peaking facilities.

irradiance. The direct, diffuse, and reflected solar radiation that strikes a surface.

J

joule. A measurement of energy. It is the work done by a force of one Newton, when the point at which the force is applied is displaced one meter in the direction of the force. It is equal to 0.239 calories. In electrical theory, one joule equals one watt-second.

K

kilovolt (kV). Equal to one thousand volts.

kilowatt-hour (kWh). A measure for energy that is equal to the amount of work done by 1000 watts for one hour. Consumers are charged for electricity in cents per kWh. One kWh is enough electricity to power a television for about three hours or to run 10, 100-watt light bulbs for one hour.

kilowatt (kW). A measurement of electric power equal to one thousand watts. Electric power capacity of 1 kW is sufficient to light 10, 100-watt light bulbs.

L

langley. A unit or measure of solar radiation; 1 calorie per square centimeter or 3.69 Btu per square foot.

liquefied natural gas. Methane that is chilled below its boiling point so it can be stored in liquid form, thereby occupying 1/625 of the space it requires at ambient temperatures and pressures.

M

magma. Molten or partially molten rock at temperatures ranging from 1260°F to 2880°F (700°C to 1600°C). Some magma bodies are believed to exist at drillable depths within the Earth's crust, although practical technologies for harnessing magma energy have not been developed. If ever utilized, magma represents a potentially enormous resource.

mass burn facility. A type of MSW incineration facility in which MSW is burned with only minor presorting to remove oversize, hazardous, or explosive materials. Mass burn facilities can be large, with capacities of 3000 tons (2.7 million kg) of MSW per day or more. They can be scaled down to handle the waste from smaller communities, and modular plants with capacities as low as 25 tons (22.7 thousand kg) per day have been built. Mass burn technologies represent more than 75% of all the MSW-to-energy facilities constructed in the United States to date. The major components of a mass burn facility include refuse receiving and handling, combustion and steam generation, flue gas cleaning, power generation (optional), condenser cooling water, residue ash hauling and landfilling.

maximum demand. The greatest of all demands of the load that has occurred within a specified period of time.

mcf. One thousand cubic feet. One Mcf of natural gas has a heating value of approximately one million Btu, also written MMBtu.

megawatt (MW). One million watts.

megawatt-hour (MWh). One million watts for one hour.

merchant plant. An electricity generating facility built and operated without long-term contracts guaranteeing sale of the electricity generated. Many such facilities are partial merchant plants, with contracts guaranteeing sale of a certain percentage of generation to a nearby utility.

methane. A colorless, odorless, tasteless gas composed of one molecule of carbon and four of H_2, which is highly flammable. It is the main constituent of *natural gas* formed naturally by methanogenic, anaerobic bacteria or can be manufactured, and which is used as a fuel and for manufacturing chemicals.

methanol (CH_3OH; methyl alcohol or wood alcohol). A clear, colorless, very mobile liquid that is flammable and poisonous, used as a fuel and fuel additive, and to produce chemicals.

methyl tertiary butyl ether (MTBE). An ether compound used as a gasoline blending component to raise the oxygen content of gasoline. MTBE is made by combining isobutylene (from various refining and chemical processes) and methanol (usually made from natural gas).

microturbine. A small gas turbine used for electric power generation.

molten carbonate fuel cells. These fuel cells run at relatively high temperatures and are very fuel flexible.

monopoly. The exclusive control of a commodity or service by one entity. In the gas industry, interstate pipelines and local distribution companies are generally monopolies. Electricity has traditionally been operated as a local monopoly. Even after deregulation, it is anticipated that transmission infrastructure will remain regulated monopolies.

municipal solid waste (MSW). Waste material from households and businesses in a community that is not regulated as hazardous.

municipal utility. An electric utility system owned and/or operated by a municipality that generates and/or purchases electricity at wholesale for distribution to retail customers generally within the boundaries of the municipality.

N

nacelle. The cover for the gearbox, drive train, generator, and other components of a wind turbine.

natural gas liquids (NGL). Hydrocarbon components of wet gas whose molecules are larger than methane but smaller than crude oil. Gas liquids include ethane, propane, and butane.

net generation. Gross generation less the electric energy consumed at the generating stations for station use.

net present value. The lifetime worth of an asset, calculated at the present time.

NIMBY. Not in my back yard.

nitrogen dioxide (NO_2). This compound of nitrogen and oxygen is formed by the oxidation of nitric oxide (NO), which is produced by the combustion of solid fuels.

nitrogen oxides (NO_x). The products of all combustion processes formed by the combination of nitrogen and oxygen.

natural gas. A naturally occurring mixture of hydrocarbon and nonhydrocarbon gases found in porous geological formations beneath the earth's surface, often in association with petroleum. The principal constituent is methane.

nonattainment area. A geographic region in the United States designated by the EPA as having ambient air concentrations of one or more criteria pollutants that exceed National Ambient Air Quality Standards.

nonutility generator (NUG). A facility that produces electric power and sells it to an electric utility, usually under long-term contract. NUGs also tend to sell thermal energy and electricity to a nearby industrial customer.

nonutility power producer. A corporation, person, agency, authority, or other legal entity that owns electric generating capacity and is not an electric utility. Nonutility power producers include qualifying small power producers and cogenerators without a designated franchised service territory.

North American Electric Reliability Council (NERC). Electric utilities formed NERC to coordinate, promote, and communicate about the reliability of their generation and transmission systems. NERC reviews the overall reliability of existing and planned generation systems, sets reliability standards, and gathers data on demand, availability, and performance.

O

ocean energy systems. Energy conversion technologies that harness the energy in tides, waves, and thermal gradients in the oceans.

ocean thermal energy conversion (OTEC). The process or technologies for producing energy by harnessing the temperature differences (thermal gradients) between ocean surface waters and that of ocean depths. Warm surface water is pumped through an evaporator containing a working fluid in a closed Rankine-cycle system. The vaporized fluid drives a turbine/generator. Cold water from deep below the surface is used to condense the working fluid. Open-cycle OTEC technologies use ocean water itself as the working fluid. Closed-cycle OTEC systems circulate a working fluid in a closed loop.

ohm. The unit of measurement of electrical resistance. Specifically, an ohm is the resistance of a circuit in which a potential difference of one volt produces a current of one Ampere.

open access. Access to the commodity market via unbundled transmission capacity, for producers, end-users, local distribution companies, and other gas resellers, on substantially equal terms for all kinds of shippers.

open-loop geothermal heat pump system. Open-loop (also known as *direct*) systems circulate water drawn from a ground or surface water source. Once the heat has been transferred into or out of the water, the water is returned to a well or surface discharge (instead of being recirculated through the system). This option is practical where there is an adequate supply of relatively clean water, and all local codes and regulations regarding groundwater discharge are met.

orientation. The alignment of a solar collector, in number of degrees east or west of true south.

outage. The period during which a generating unit, transmission line, or other facility is out of service.

oxygenates. Gasoline fuel additives such as ethanol, ETBE, or MTBE that add extra oxygen to gasoline to reduce CO pollution produced by vehicles.

ozone transport. Ozone transport occurs when emissions from one area drift downwind and mix with local emissions contributing to the ozone concentrations in the downwind area.

ozone. A compound consisting of three oxygen atoms. It is the primary constituent of smog.

P

panel (solar). A term generally applied to individual solar collectors, and typically to solar PV collectors or modules.

panel radiator. A mainly flat surface for transmitting radiant energy.

panemone. A drag-type wind machine that can react to wind from any direction.

parabolic dish. A solar energy conversion device that has a bowl shaped dish covered with a highly reflective surface that tracks the sun and concentrates sunlight on a fixed absorber, thereby achieving high temperatures, for process heating or to operate a heat (Stirling) engine to produce power or electricity.

parabolic trough. A solar energy conversion device using trough covered with a highly reflective surface to focus sunlight onto a linear absorber containing a working fluid that can be used for medium temperature space or process heat or to operate a steam turbine for power or electricity generation.

parallel connection. A way of joining PV cells or modules by connecting positive leads and negative leads together; such a configuration increases the current, but not the voltage.

passive solar (building) **design.** A building design that uses structural elements of a building to heat and cool a building, without the use of mechanical equipment, which requires careful consideration of the local climate and solar energy resource, building orientation, and landscape features, to name a few. The principal elements include proper building orientation, proper window sizing, and placement and design of window overhangs to reduce summer heat gain and ensure winter heat gain, and proper sizing of thermal energy storage mass (e.g., a Trombe wall or masonry tiles). The heat is distributed primarily by natural convection and radiation, though fans can also be used to circulate room air or ensure proper ventilation.

passive solar heater. A solar water or space-heating system in which solar energy is collected, and/or moved by natural convection without using pumps or fans. Passive systems are typically integral collector/storage (ICS; or batch collectors) or thermosyphon systems. The major advantage of these systems is that they do not use controls, pumps, sensors, or other mechanical parts, so little or no maintenance is required over the lifetime of the system.

peak days. In electricity, the days in the summer months when the demand for electricity is at its highest level due to air-conditioning load. For natural gas, peak days are the days in the winter months when demand for gas is at its highest level due to most heating equipment being used.

peak load plant. A plant usually housing old, low-efficiency steam units; gas turbines; diesels; or pumped-storage hydroelectric equipment normally used during the peak-load periods.

peak sun hours. The equivalent number of hours per day when solar irradiance averages 1 kW/m^2.

peak wind speed. The maximum instantaneous wind speed (or velocity) that occurs within a specific period of time or interval.

peaking capacity. Capacity of generating equipment normally reserved for operation during the hours of highest daily, weekly, or seasonal loads. Some generating equipment may be operated at certain times as peaking capacity and at other times to serve loads on an around-the-clock basis.

pellets. Solid fuels made from primarily wood sawdust that is compacted under high pressure to form small (about the size of rabbit feed) pellets for use in a pellet stove.

pellet stove. A space heating device that burns pellets; is more efficient, clean burning, and easier to operate relative to conventional cord wood burning appliances.

pelton turbine. A type of impulse hydropower turbine where water passes through nozzles and strikes cups arranged on the periphery of a runner, or wheel, which causes the runner to rotate, producing mechanical energy. The runner is fixed on a shaft, and the rotational motion of the turbine is transmitted by the shaft to a generator. Generally used for high head, low flow applications.

penstock. A component of a hydropower plant; a pipe that delivers water to the turbine.

phosphoric acid fuel cells. The only commercially available type of fuel cell; these offer very high efficiencies in cogeneration uses. They have a slow start-up time so are not suited to emergency power needs.

photoelectric cell. A device for measuring light intensity that works by converting light falling on, or reaching it, to electricity, and then measuring the current; used in photometers.

photoelectrochemical cell. A type of PV device in which the electricity induced in the cell is used immediately within the cell to produce a chemical, such as H_2, which can be withdrawn for use.

photon. A particle of light that acts as an individual unit of energy.

photovoltaic (PV) efficiency. The ratio of the electric power produced by a PV device to the power of the sunlight incident on the device.

photovoltaic array. A group of solar PV modules connected together.

photovoltaic cell. Treated semiconductor material that converts solar irradiance to electricity.

photovoltaic device. A solid-state electrical device converting light directly into direct current electricity of voltage-current characteristics that are a function of the characteristics of the light source and the materials in and design of the device. Solar PV devices are made of various semiconductor materials including silicon, cadmium sulfide, cadmium telluride, and gallium arsenide, and in single crystalline, multicrystalline, or amorphous forms.

photovoltaic module or panel. A solar PV product generally consisting of groups of PV cells electrically connected together to produce a specified power output under standard test conditions, mounted on a substrate, sealed with an encapsulant, and covered with a protective glazing. Maybe further mounted on an aluminum frame. A junction box, on the back or underside of the module is used to allow for connecting the module circuit conductors to external conductors.

photovoltaic system. A complete PV power system composed of the module (or array), and balance-of-system (BOS) components including the array supports, electrical conductors/wiring, fuses, safety disconnects, and grounds, charge controllers, inverters, battery storage, etc.

pitch control. A method of controlling a wind turbine's speed by varying the orientation, or pitch, of the blades, and thereby altering its aerodynamics and efficiency.

plant-use electricity. The electric energy used in the operation of a plant. This energy total is subtracted from the gross energy production of the plant; for reporting purposes the plant energy production is then reported as a net figure. The energy required for pumping at pumped-storage plants is, by definition, subtracted, and the energy production for these plants is then reported as a net figure.

polymer electrolyte membrane fuel cells. Also called proton exchange membrane fuel cells. These are typically developed for smaller applications, such as light-duty vehicles, small buildings, and for electronics such as video cameras and laptop computers.

portfolio standard. The requirement that an electric power provider generate or purchase a specified percentage of the power it supplies/sells from renewable energy resources, and thereby guarantee a market for electricity generated from renewable energy resources.

power. The instantaneous current being delivered at a given voltage, measured in watts, or more usually kw. Power delivered for a period of time is energy, measured in kWh.

power generation mix. The proportion of electricity distributed by a power provider that is generated from available sources such as coal, natural gas, petroleum, nuclear, hydropower, wind, or geothermal.

power tower. A term used to describe solar thermal, central receiver, power systems, where an array of reflectors focus sunlight onto a central receiver and absorber mounted on a tower.

prime mover. The engine, turbine, water wheel, or similar machine that drives an electrical generator. Generally, a prime mover refers to a device that converts energy to electricity directly, such as PV solar and fuel cells.

private power producer. Any entity engaging in wholesale power generation or in self-generation.

privatization. The sale or transfer to private individuals or businesses of assets or businesses owned by the government or the conversion of a government-owned firm or industry to private ownership.

pro-forma. A general budget describing the basic economic expectations of a venture.

propeller turbine (hydro). A turbine that has a runner with attached blades similar to a propeller used to drive a ship. As water passes over the curved propeller blades, it causes rotation of the shaft.

public utility. Publicly owned electric utilities are nonprofit local government agencies established to serve their communities and nearby consumers at cost, returning excess funds to the consumer in the form of community contributions, economic and efficient facilities, and lower rates. Publicly owned electric utilities number approximately 2000 in the United States, and include municipals, public power districts, state authorities, irrigation districts, and others.

Public Utility Commission (PUC). An administrative or quasi-judicial body at the state provincial or municipal level, whose functions include regulation of public utilities.

Public Utility Holding Company Act of 1935 (PUHCA). PUHCA regulated the large interstate holding companies that monopolized the electric utility industry in the early part of the twentieth century. Recently, the Securities and Exchange Commission has been interpreting this legislation more leniently, allowing foreign firms to buy domestic utilities and allowing nonutility companies to purchase utilities without becoming registered holding companies. It is broadly believed that as deregulation of the electric industry becomes more comprehensive, PUHCA will either be repealed or replaced.

Public Utility Regulatory Policy Act of 1978 (PURPA). PURPA promotes energy efficiency and increased use of alternative energy sources, encouraging companies to build cogeneration facilities and renewable energy projects. Facilities meeting PURPA requirements are called qualifying facilities or QFs.

pumped storage facility. A type of power generating facility that pumps water to a storage reservoir during off-peak periods, and uses the stored water (by allowing it to fall through a hydro turbine) to generate power during peak periods. The pumping energy is typically supplied by lower cost base power capacity, and the peaking power capacity is of greater value, even though there is a net loss of power in the process.

pyranometer. A device used to measure total incident solar radiation (direct beam, diffuse, and reflected radiation) per unit time per unit area.

pyrheliometer. A device that measures the intensity of direct beam solar radiation.

pyrolysis. Pyrolysis of biomass is the thermal degradation of the material in the absence of reacting gases, and occurs prior to or simultaneously with gasification reactions in a gasifier. Pyrolysis products consist of gases, liquids, and char generally. The liquid fraction of pyrolisized biomass consists of an insoluble viscous tar, and pyroligneous acids (acetic acid, methanol, acetone, esters, aldehydes, and furfural). The distribution of pyrolysis products varies depending on the feedstock composition, heating rate, temperature, and pressure.

Q

quad. Abbreviation for one quadrillion Btu. For natural gas, this is roughly one trillion cubic feet.

qualifying facility (QF). A generator that (1) qualifies as a cogenerator or small power producer under PURPA, and (2) has obtained certification from FERC. They generally sell power to utilities at the utilities' avoided cost.

R

Rankine cycle. The thermodynamic cycle that is an ideal standard for comparing performance of heat-engines, steam power plants, steam turbines, and heat pump systems that use a condensable vapor as the working fluid; efficiency is measured as work done divided by sensible heat supplied.

recuperator. A heat exchanger in which heat is recovered from the products of combustion.

reflectance. The amount (percent) of light that is reflected by a surface relative to the amount that strikes it.

refuse-derived fuel (RDF). A solid fuel produced by shredding MSW. Noncombustible materials such as glass and metals are generally removed prior to making RDF. The residual material is sold as is or compressed into pellets, bricks, or logs. RDF processing facilities are typically located near a source of MSW, while the RDF combustion facility can be located elsewhere.

regulation. The government function of controlling or directing economic entities through the process of rule-making and adjudication.

regulatory environment. Regulations and enforcement affecting marketing activities laid down by government and nongovernment entities.

renewable energy. Refers to any source of energy that is constantly replenished through natural processes. Sunlight, moving water, geothermal springs, biomass, and wind are all examples of renewable energy resources used to generate electricity.

run-of-river hydropower. A type of hydroelectric facility that uses the river flow with very little alteration and little or no impoundment of the water.

S

silicon. A chemical element, of atomic number 14, that is semimetallic, and an excellent semiconductor material used in solar PV devices; commonly found in sand.

solar altitude angle. The angle between a line from a point on the earth's surface to the center of the solar disc, and a line extending horizontally from the point.

solar array. A group of solar collectors or solar modules connected together.

solar azimuth. The angle between the sun's apparent position in the sky and true south, as measured on a horizontal plane.

solar cell. A solar PV device with a specified area.

solar collector. A device used to collect, absorb, and transfer solar energy to a working fluid. Flat plate collectors are the most common type of collectors used for solar water or pool heating systems. For example, in the case of a PV system, the solar collector could be crystalline silicon panels or thin-film roof shingles.

solar constant. The average amount of solar radiation that reaches the earth's upper atmosphere on a surface perpendicular to the sun's rays; equal to 1353 watts per square meter or 492 Btu per square foot.

solar declination. The apparent angle of the sun north or south of the earth's equatorial plane. The earth's rotation on its axis causes a daily change in the declination.

solar energy. Electromagnetic energy transmitted from the sun (solar radiation). The amount that reaches the earth is equal to one billionth of total solar energy generated, or the equivalent of about 420 trillion kWh.

Solar Energy Industries Association (SEIA). A national trade association of solar energy equipment manufacturers, retailers, suppliers, installers, and consultants.

Solar Energy Research Institute (SERI). A federally funded institute, created by the Solar Energy Research, Development and Demonstration Act of 1974, that conducted research and development of solar energy technologies. Became the NREL in 1991.

solar irradiation. The amount of solar radiation, both direct and diffuse, received at any location.

solar noon. The time of the day, at a specific location, when the sun reaches its highest, apparent point in the sky; equal to true or due, geographic south.

solar radiation. A general term for the visible and near visible (ultraviolet and near-infrared) electromagnetic radiation that is emitted by the sun. It has a spectral, or wavelength, distribution that corresponds to different energy levels; short-wavelength radiation has a higher energy than long-wavelength radiation.

solar thermal electric systems. Solar energy conversion technologies that convert solar energy to electricity, by heating a working fluid to power a turbine that drives a generator. Examples of these systems include central receiver systems, parabolic dish, and solar trough.

solar thermal parabolic dishes. A solar thermal technology that uses a modular mirror system approximating a parabola and incorporating two-axis tracking to focus the sunlight onto receivers located at the focal point of each dish. The mirror system typically is made from a number of mirror facets, either glass or polymer mirror, or can consist of a single stretched membrane using a polymer mirror. The concentrated sunlight may be used directly by a Stirling, Rankine, or Brayton cycle heat engine at the focal point of the receiver or to heat a working fluid that is piped to a central engine. The primary applications include remote electrification, water pumping, and grid-connected generation.

solid oxide fuel cell. These can be differentiated from other fuel cell types by their high operating temperature and solid-state ceramic cell structure. These may someday be used in combined-cycle fuel cell hybrid applications with high efficiencies.

split spectrum PV cell. A PV device where incident sunlight is split into different spectral regions, with an optical apparatus, that is directed to individual PV cells optimized for converting that spectrum to electricity.

steam electric plant. A plant in which the prime mover is a steam turbine. The steam used to drive the turbine is produced in a boiler where fossil fuels are burned.

storage hydropower. A hydropower facility that stores water in a reservoir during high-inflow periods to augment water during low-inflow periods. Storage projects allow the flow releases and power production to be more flexible and dependable. Many hydropower project operations use a combination of approaches.

stranded costs. This refers to a utility's fixed costs, usually related to investments in generation facilities, which would no longer be paid by customers through their rates in the event they opt to purchase power from other suppliers.

T

temperature coefficient (of a solar PV cell). The amount that the voltage, current, and/or power output of a solar cell changes due to a change in the cell temperature.

therm. One therm equals 100,000 Btu.

thin-film. A layer of semiconductor material, such as copper indium diselenide or gallium arsenide, a few microns or less in thickness, used to make solar PV cells.

tidal power. The power available from the rise and fall of ocean tides. A tidal power plant works on the principal of a dam or barrage capturing water in a basin at the peak of a tidal flow, then directing the water through a hydroelectric turbine as the tide ebbs.

tilt angle (of a solar collector or module). The angle at which a solar collector or module is set to face the sun relative to a horizontal position. The tilt angle can be set or adjusted to maximize seasonal or annual energy collection.

time-of-day pricing. A rate structure that prices electricity at different rates, reflecting the changes in the utility's costs of providing electricity at different times of the day.

tracking solar array. A solar energy array that follows the path of the sun to maximize the solar radiation incident on the PV surface. The two most common orientations are (1) one axis where the array tracks the sun east to west, and (2) two-axis tracking where the array points directly at the sun at all times. Tracking arrays use both the direct and diffuse sunlight. Two-axis tracking arrays capture the maximum possible daily energy.

transmission grid. The high voltage wires that connect generation facilities with distribution facilities. It is the infrastructure through which power moves around the United States. It is necessary to carefully coordinate use of the transmission system to ensure reliable and efficient service.

transmission line. A set of conductors, insulators, supporting structures, and associated equipment used to move large quantities of power at high voltage.

transmission system. An interconnected group of electric transmission lines and associated equipment for moving or transferring electric energy in bulk between points of supply and points at which it is transformed for delivery over the distribution lines to consumers or is delivered to other electric systems.

transmission. The movement or transfer of electric energy over an interconnected group of lines and associated equipment between points of supply and points at which it is transformed for delivery to consumers, or to other electric systems. Transmission is considered to end when the energy is transformed for distribution to the consumer; in natural gas, the conveyance of natural gas from producing to consuming areas through large-diameter, high-pressure pipelines.

turbine. A machine for generating rotary mechanical power from the energy of a stream of fluid (e.g., water, steam, or hot gas).

U

ultra-high voltage systems. Electric systems in which the operating voltage levels have a maximum root-mean-square AC voltage above 800,000 volts (800 kV).

utility. Privately owned companies and public agencies engaged in the generation, transmission, or distribution of electric power for public use.

V

value-added services. Services, such as security monitoring, telecommunications, internet access, and others, that add value to electric services. Other services which can be offered by utilities to achieve greater customer satisfaction and loyalty.

variable-speed wind turbines. Turbines in which the rotor speed increases and decreases with changing wind speed, producing electricity with a variable frequency.

vertical-axis wind turbine (VAWT). A type of wind turbine in which the axis of rotation is perpendicular to the wind stream and the ground.

volt. The measure of pressure that pushes electric current through a circuit.

volt-ampere reactive (VAR). A reactive load, typically inductive from electric motors, which causes more current to flow in the distribution network than is actually consumed by the load. This requires excess capability on the generation side and causes greater power losses in the distribution network.

W

waste-to-energy plants. A steam-turbine generating facility that uses MSW as the primary energy source to produce the steam used in the generating process.

water turbine. A turbine that uses water pressure to rotate its blades; the primary types are the Pelton wheel, for high heads (pressure); the Francis turbine, for low to medium heads; and the Kaplan for a wide range of heads. Primarily used to power an electric generator.

water wheel. A wheel that is designed to use the weight and/or force of moving water to turn it, primarily to operate machinery or grind grain.

watt. The basic expression of electrical power or the rate of electrical work. One watt is the power resulting from the dissipation of one joule of energy in one second.

watt-hour. An electrical energy unit of measure equal to one watt of power supplied to, or taken from, an electric circuit steadily for one hour.

wheeling. The transportation of power to customers. Wholesale wheeling is transmitting bulk power over the grid to power companies. Retail wheeling is transmitting power to end users (e.g., homes, businesses, and factories).

wind energy. Energy available from the movement of the wind across a landscape caused by the heating of the atmosphere, the earth, and oceans by the sun.

windpower curve. A graph representing the relationship between the power available from the wind and the wind speed. The power from the wind increases proportionally with the cube of the wind speed.

windpower plant. A group of wind turbines interconnected to a common power provider system through a system of transformers, distribution lines, and usually one substation. Operation, control, and maintenance functions are often centralized through a network of computerized monitoring systems, supplemented by visual inspection. This is a term commonly used in the United States. In Europe, it is called a generating station.

windpower profile. The change in the power available in the wind due to changes in the wind speed or velocity profile; the windpower profile is proportional to the cube of the wind speed profile.

wind resource assessment. The process of characterizing the wind resource, and its energy potential, for a specific site or geographical area.

wind speed. The rate of flow of the wind undisturbed by obstacles.

wind speed duration curve. A graph that indicates the distribution of wind speeds as a function of the cumulative number of hours that the wind speed exceeds a given wind speed in a year.

wind speed frequency curve. A curve that indicates the number of hours per year that specific wind speeds occur.

wind speed profile. A profile of how the wind speed changes with height above the surface of the ground or water.

wind turbine. A term used for a wind energy conversion device that produces electricity; typically having one, two, or three blades.

wind turbine rated capacity. The amount of power a wind turbine can produce at its rated wind speed (e.g., 100 kW at 20 mph). The rated wind speed generally corresponds to the point at which the conversion efficiency is near its maximum. Because of the variability of the wind, the amount of energy a wind turbine actually produces is a function of the capacity factor (e.g., a wind turbine produces 20% to 35% of its rated capacity over a year).

wind velocity. The wind speed and direction in an undisturbed flow.

Y

yaw. The rotation of a horizontal axis wind turbine around its tower or vertical axis.

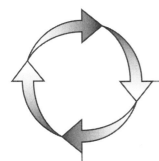

Index

A

Address list (industry contacts), 197–203:
  domestic, 197–202;
  international, 202–203
Advanced Thermal Systems, Inc., 139–142
AESE engineering firm (solar/PV case study), 32–34:
  system specifications, 33–34
Aesthetic concerns (wind power), 78–79, 82–85:
  offshore, 82–85
Affordability (PV), 29
Africa, 16–17
Air ventilation, 25
Alabama Power (bioenergy case study), 118–120
Albedo radiation (solar), 39
Alkaline fuel cell, 180–181
Ambient ground heat, 130
American Biofuels Association, 116
American Electric Power, 49–50
American Wind Energy Association (AWEA), 69, 71
Americans with Disabilities Act of 1990 (ADA), 169
Argentina, 15
Arizona Public Service (APS), 27, 54–57
Arklow Bank project, Ireland (wind energy case study), 93–95
Asia, 13–14:
  industrialized Asia, 13;
  developing Asia, 14
AstroPower solar power/PV system, 30–31
At-Home Services (Home Depot), 30–31

Australasia, 13

Australia solar power tower
(solar power/air flow case study), 58–60:
financing, 59;
design and construction, 60

Australia, 13, 58–60, 67:
power tower (case study), 58–60

B

Bangladesh (bioenergy case study), 126–128

Belgium, 85

Big Spring, Tex.
(wind energy case study), 80–82

Biobased products, 120–122

Biodiesel, 98, 115–120:
market, 116–117;
Savannah River Site (case study), 117–118;
Florida Power & Light
(case study), 118–120;
Alabama Power (case study), 118–120

Biodigester, 127–128

Bioenergy/biomass energy, 5, 7–8, 97–128:
methanol, 98;
biodiesel, 98, 115–120;
reformulated gasoline components, 98;
dedicated energy crops, 99;
landfill gas-to-energy, Iowa
(case study), 99–100;
electricity generation, 101;
biomass potential, 102–105;
Portland General Electric
(case study), 106–107;
District Energy St. Paul heat and power plant, Minnesota (case study), 107–108;
Energy Products of Idaho power plant
(case study), 108–109;
biogas potential, 109–111;
biopower basics, 111–114;
transportation applications, 114;
ethanol, 114–115;

Savannah River Site (case study), 117–118;
Florida Power & Light
(case study), 118–120;
Alabama Power (case study), 118–120;
other biobased products, 120–122;
biomass resources, 122–124;
world views, 124–125;
Bangladesh (case study), 126–128

Biogas, 109–111, 113:
potential, 109–111

Biomass briquetting/densification, 127

Biomass energy.
SEE Bioenergy/biomass energy.

Biomass fuel processes, 101, 112–114:
direct combustion, 101, 112–113;
gasification, 101, 113;
pyrolysis, 101, 113;
co-firing, 113;
modular systems, 114

Biomass potential, 102–105

Biomass resources, 122–124

Biopower basics, 101, 111–114:
direct combustion, 101, 112–113;
gasification, 101, 113;
pyrolysis, 101, 113;
co-firing, 113;
modular systems, 114

Biosynthesis gas, 121

Bird problem (wind energy), 72, 76–77

Blade shape (wind turbine), 88

BP Solar Latin America, 44–45

Brazil, 15, 95–96, 150, 155–157:
wind energy case study, 95–96;
hydroelectric/water power
case study, 155–157

Briquetting/densification (biomass), 127

British Petroleum Solar, 31–32

Building-integrated PV (BIPV), 36, 44

C

Cal-Gon Farms, 106–107

California Public Utility Commission (PUC), 42

California State University Northridge (solar/PV case study), 45–46

Calpine and the Geysers field (geothermal energy case study), 142–144

Canada, 148, 151–152:
hydroelectric/water power, 152–154

Capacity, 86–88, 135:
wind energy, 86–88;
capacity factor, 135

Cape Wind Associates project, 82–83

Carbon dioxide, 43, 109, 112, 114, 134–135, 163, 178–179:
emission, 43, 109, 112, 114, 134–135, 163

Carbon monoxide, 121, 179–180

Carbonate fuel cell, 191–192

Case studies (bioenergy/biomass energy), 99–100, 106–109, 117–120, 126–128:
landfill gas-to-energy (Iowa), 99–100;
Portland General Electric (Oregon), 106–107;
District Energy St. Paul heat and power plant (Minnesota), 107–108;
Energy Products of Idaho, 108–109;
Savannah River Site (DOE), 117–118;
Florida Power & Light, 118–120;
Alabama Power, 118–120;
Bangladesh, 126–128

Case studies (fuel cells), 191–192:
Wabash River energy facility, Indiana, 191–192

Case studies (geothermal energy), 139–144:
University of Nevada, Reno, 139–142;
Calpine and the Geysers field, 142–144

Case studies (hydroelectric/water power), 155–157, 170–171:
Brazil energy crisis, 155–157;
Hawaii, 170–171

Case studies (solar energy/PV), 32–34, 36–38, 41–60:
AESE engineering firm, 32–34;
Solar Patriot house green design, 36–38;
Santa Rita Jail, 41–42;
United Kingdom, 42–44;
Peru telecommunication systems, 44–45;
California university solar panels, 45–46;
Peru solar grid-connect system, 47;
Pacific Park, Calif., 48–49;
Learning From Light program, 49–50;
Parker Ranch, Hawaii (solar/wind hybrid), 50–53;
Glendale, Ariz. High concentration PV, 54–57;
Australia solar power tower (solar/air flow), 58–60

Case studies (wind energy), 80–82, 91–96:
Big Spring, Tex., 80–82;
Italy, 91–93;
Arklow Bank project, Ireland, 93–95;
Brazil, 95–96

Central America, 15, 150, 155:
hydroelectric/water power, 155

China, 14, 67, 125

City Public Service, San Antonio, Tex., 26

Clean Water Act of 1977 (CWA), 168

Coal fuel, 2, 6–7, 16–17, 103, 109, 112

Coastal Zone Management Act of 1972 (CZMA), 169

Cogeneration/co-firing, 8, 103, 113, 122–125

Combined heat and power (CHP) system, 183–184

Comisión Federal de Electricidad (CFE), 11

Commercialization (fuel cells), 182–184

Congo (Kinshasa), 16

Contacts (industry addresses), 197–203:
  domestic, 197–202;
  international, 202–203
Conventional hydropower, 158
Costa Rica, 15
Crystalline silicon solar cell, 41
Cut-in/cutout speed (wind turbine), 88

D

Darrius model wind turbine, 62
Database of State Incentive for Renewable Energy (DSIRE), 9
Daylighting (passive solar heating), 23–24:
  sunspace, 24;
  trombe wall, 24
Dedicated energy crops, 99, 103, 123–124
Definitions, 205–232
Denmark, 13, 66, 85
Development forecast (hydroelectric/water power), 164–165
Diffuse radiation (solar), 39
Direct combustion (biomass), 101, 112–113
Direct fuel cell technology, 191–192
Direct radiation (solar), 39
Dish/engine system, 22–23
District Energy St. Paul heat and power plant, Minn. (bioenergy case study), 107–108
Diversion (hydroelectric/water power), 160

E

Eastern Europe, 16
Eco Solar, 47
Efficiency measures (high concentration PV), 57

Egypt, 16–17
Electric Power Research Institute (EPRI), 27
Electricity, 1–7, 15, 21–22, 88–91:
  U.S. transmission grid, 88–91
Emerging Renewable Buydown program, 42
Endangered Species Act of 1973 (ESA), 169
Energy crops, 99, 103, 123–124
Energy Policy Act (EPAct), 116–117, 119
Energy Products of Idaho (bioenergy case study), 108–109
Enron Wind Corp., 71
EnviroMission, 59
Environmental issues (hydroelectric/water power), 159, 163:
  fishway/fish ladder, 159;
  fish mortality, 163;
  water quality standards, 163;
  carbon dioxide emissions, 163
Environmental issues (wind energy), 72, 76–79, 82–85:
  bird problem, 72, 76–77;
  sound, 72, 77–78;
  aesthetics, 78–79, 82–85
ETBE, 98, 120
Ethanol, 114–115
Europe, 12–13, 16, 66–67, 188–189, 193–194:
  Western Europe, 12–13;
  Eastern Europe, 16;
  fuel cell market, 193–194
European Union (EU), 12, 66–67, 189
European Wind Energy Association (EWEA), 70
Evacuated-tube collector (solar power/PV)

F

Federal Power Act (FPA), 166–167
FERCO, 104–105

Fish and Wildlife Coordination Act of 1934 (FWCA), 168

Fish mortality, 163

Fishway/fish ladder, 159

Fitel project, 44–45

Florida Power & Light (bioenergy case study), 118–120

Fluidized bed combustion system, 109–111

Former Soviet Union, 16

Fossil fuels, 2, 6–7, 16–17, 22

France, 12

Francis turbine, 161

Fresnel lens, 56–57

Fuel cell heating appliance (FCHA) design, 193

Fuel cell types, 176–182:
 phosphoric acid fuel cell, 176–178, 182;
 molten carbonate fuel cell, 176, 178–179, 182;
 solid oxide fuel cell, 176, 179–180, 182;
 polymer exchange membrane fuel cell, 180–182;
 alkaline fuel cell, 180–181

Fuel cells, 173–195:
 fuel cell types, 176–182;
 commercialization, 182–184;
 popularity, 184–186;
 capital costs, 184–186, 188;
 reliability, 187;
 hydrogen economy, 187–188;
 geography, 188–190;
 future outlook, 190–191;
 Wabash River energy facility, Indiana (case study), 191–192;
 Europe residential market, 193–194;
 hydrogen issues, 195

Fuel type, 1–2, 6–7

FuelCell Energy (FCE), 185–186, 191

Fullerene solar cell, 41

Funding strategies (solar/PV), 29

G

Gardner, Massachusetts, 27

Gasification (biomass), 101, 113

Gasoline, 98, 114–115, 120:
 reformulated gasoline components, 98, 120

GE Wind Energy, 69–70

GeoPowering the West, 139

Geopressured brines, 130, 136

Geothermal Energy Program (DOE), 138–139

Geothermal energy, 5, 7–8, 129–146:
 steam, 131;
 high-temperature water, 131;
 moderate-temperature water, 132–133;
 power production, 133–135;
 future of geothermal power, 135–136, 144–146;
 DOE developments, 137;
 research and development, 137–139;
 University of Nevada, Reno (case study), 139–142;
 Calpine and the Geysers field (case study), 142–144;
 outlook, 144–146

Geothermal resources, 144–146

Germany, 12, 66, 85

Geysers geothermal field, 132, 142–145:
 geothermal energy case study, 142–144

Glendale, Ariz. High concentration PV (solar/PV case study), 54–57:
 how it works, 56–57;
 efficiency measures, 57

Global positioning system (GPS), 57

Glossary, 205–232

Green certificates, 12–13

Green design (house), 36–38

Green Power I project, 71

Green power, 1–2

Green premium, 110

Grid-connected solar/PV, 28, 43–44, 47

Guerilla Solar, 27

Guidelines (wind energy), 73–74

H

Hawaii (hydroelectric/water power case study), 170–171:
Kauai, 170–171;
Maui, 171;
Island of Hawaii, 171

Head (hydropower), 157

Heat and power plant, Minnesota (bioenergy case study), 107–108

Heating, ventilating, and air conditioning (HVAC), 37

High concentration PV (case study), 54–57:
how it works, 56–57;
efficiency measures, 57

High-temperature water, 131

High-wind scenarios, 91

Holland, 12–13, 85

Home Depot At-Home Services, 30–31

Horizontal-axis wind turbine, 62

Hot dry rock, 130, 136

Hybrid solar/wind case study, 50–53

Hydroelectric/water power types, 160–162:
impoundment, 160;
diversion, 160;
storage facilities, 160;
plant size, 161;
turbine technologies, 161–162

Hydroelectric/water power, 5–7, 13–17, 147–171:
North America, 151;
United States, 151–152;
Canada, 152–154;
Central America, 155;
South America, 155;
Brazil energy crisis (case study), 155–157;
hydropower basics, 157–160;
hydropower types, 160–162;
environmental issues and mitigation, 163;
development forecast, 164–165;
relicensing, 166–167;
legislative and regulatory considerations, 167–169;
Hawaii (case study), 170–171

Hydrogen economy (fuel cells), 187–188

Hydrogen, 121, 174–175, 187–188, 195:
hydrogen economy, 187–188;
fuel cells, 195;
safety, 195

Hydropower basics, 157–160:
conventional, 158;
pumped storage, 158–160

Hydro-Quebec, 154

Hydrothermal fluids, 130–131

I

Iceland, 146

Impoundment (hydroelectric/water power), 160

India, 14, 66–67, 125

Indonesia, 145

Industry contact list, 197–203:
domestic, 197–202;
international, 202–203

Institute of Electrical and Electronic Engineers (IEEE), 89

Integrated gasification fuel cell, 192

Interconnected grid (solar/PV), 28, 43–44, 47

International Clean Energy Initiative, 139

International industry contacts, 202–203

Iran, 16

Ireland, 85, 93–95:
  wind power case study, 93–95
Italy, 13, 80–82, 91–93, 145:
  wind power case study, 80–82, 91–93
Ivory Coast, 16

J

Japan, 13, 67

K

Kalina cycle technology, 139–142
Kaplan turbine, 162
Kauai, 170–171
Kenya, 16

L

Land use (wind energy), 74–75
Landfill gas, 4, 7, 99–100:
  case study, 99–100
Landfill gas-to-energy, Iowa
  (bioenergy case study), 99–100
Learning From Light program
  (solar/PV case study), 49–50
Legislative/regulatory considerations
  (hydroelectric/water power), 167–169:
  National Environmental Policy Act of 1969,
    167–168;
  Clean Water Act of 1977, 168;
  Fish and Wildlife Coordination
    Act of 1934, 168;
  National Historic Preservation
    Act of 1966, 168;
  Wild and Scenic Rivers
    Act of 1968, 168–169;
  Endangered Species
    Act of 1973, 169;
  Coastal Zone Management
    Act of 1972, 169;
  Americans with Disabilities
    Act of 1990, 169
Life-cycle cost basis, 64–65

M

Madera Power, 108–109
Magma, 130, 136
Malaysia, 14
Manufacturing (solar power/PV), 29
Maui, 171
Mauna Lani Hotel, 27
McNeil Generation Station, 104–105
Merchant potential (wind energy), 71
Mercury, 109
Methane gas, 99–100, 113, 127–128
Methanol, 98, 120
Metro Park East Sanitary Landfill,
  Des Moines, Iowa, 99–100
Mexico, 11
Middle East, 16–17
Moderate-temperature water, 132–133
Modular systems (bioenergy), 114
Molten carbonate fuel cell
  (MCFC), 176, 178–179, 182
Morocco, 16
MTBE, 98, 120
Municipal solid waste
  (MSW), 3, 5, 115, 127–128

N

Nano-crystalline solar cell, 41

National Aeronautics and Space Administration (NASA), 173–174

National Environmental Policy Act of 1969 (NEPA), 167–168

National Fuel Cell Research Center, 179

National Historic Preservation Act of 1966 (NHPA), 168

National Hydropower Association, 164–165

National Renewable Energy Laboratory (NREL), 22, 35

Natural gas, 2, 6–7, 15–16, 95, 103, 150, 190

Net metering, 47

Netherlands, 12–13, 85

Nevada Renewable Energy and Energy Conservation Task Force, 142

New Zealand, 13, 145–146

Niagara Mohawk, 27

Nitrogen oxide, 109

Noise impacts (wind power), 72, 77–78

Nonenergy added value (PV), 28

Nonsilicon compound thin film solar cell, 41

North America (hydroelectric/water power), 151

Nuclear power, 2, 13, 17

O

Office National d' Electricité, 16

Offshore wind power installations, 82–86:
aesthetics, 82–85;
capital costs, 85–86

On-grid renewables, 151

OptiSlip control system, 82

Output (wind turbine), 86–88:
cut–in speed, 88;
blade shape, 88;
cutout speed, 88;
operating characteristics, 88;
efficiency, 88

P–Q

Pacific Park, Calif. (solar/PV case study), 48–49

Pacific Wheel, 48–49

Parabolic-trough system, 22–23, 25

Parker Ranch, Hawaii (solar/wind case study), 50–53:
equipment, 52;
benefits, 52;
future views, 52–53

Passive solar heating and daylighting, 23–25:
sunspace, 24;
trombe wall, 24

Peak shaving, 33

Pelton turbine, 161

Penstocks (hydropower), 158

Permitting issues (wind power), 73–74

Peru telecommunication systems (solar/PV case study), 44–45

Phenol, 120

Phosphoric acid fuel cell (PAFC), 176–178, 182

Photovoltaics (PV), 19–60:
solar power, 19–20;
solar basics, 21–32; AESE engineering firm (case study), 32–34;
DOE Solar Buildings Program, 34–35;
Solar Patriot house green design (case study), 36–38;
solar radiation types, 39;
solar factoids, 39;

solar cell types, 40–41;
Santa Rita Jail (case study), 41–42;
United Kingdom (case study), 42–44;
Peru telecommunication systems
 (case study), 44–45;
California university solar panels
 (case study), 45–46;
Peru solar grid-connect system
 (case study), 47;
Pacific Park, Calif. (case study), 48–49;
Learning From Light program
 (case study), 49–50;
Parker Ranch, Hawaii (case study), 50–53;
Glendale, Ariz. High concentration PV
 (case study), 54–57;
Australia solar power tower plant
 (case study), 58–60

Plant size (hydroelectric/water power), 161

Plug Power, 183, 194

Polymer electrolyte fuel cell (PEFC), 181–182

Polymer exchange membrane
 (PEM) fuel cell, 180–182:
 polymer electrolyte fuel cell, 181–182

Portland General Electric
 (bioenergy case study), 106–107

Power tower, 22–23, 58–60:
 basic concept, 23;
 Australia (case study), 58–60;
 financing, 59;
 design and construction, 60

PowerGuard solar tile, 42

PowerLight Corp., Berkeley, Calif., 41–42

Production increase (PV), 29

Propeller turbine, 162

Proton exchange membrane fuel cell, 182

Pumped storage (hydropower), 158–160

PV arrays, 21–22

PV market, 26–29:
 Sacramento Municipal Utility District, 27;
 Shea Homes, 27;
 Niagara Mohawk, 27;
 Gardner, Mass., 27;
 Arizona Public Service, 27;
 Mauna Lani Hotel, 27;
 Stelle, Ill., 27;
 Guerilla Solar, 27;
 grid-connected PV, 28;
 nonenergy added value, 28;
 grid interconnection, 28;
 reliability, 28;
 location, 29;
 strategies for funding, 29;
 teaming, 29;
 affordability, 29;
 increased production, 29

Pyrolysis (biomass), 101, 113

R

Radiation (solar), 39

Rated power (wind turbine), 86–88

Reformulated gasoline components, 98, 120

Regulations.
 SEE Legislative/regulatory considerations
 (hydroelectric/water power).

Reliability issues, 28, 187:
 solar/PV, 28;
 fuel cells, 187

Relicensing
 (hydroelectric/water power), 166–167

Renewable energy sources, 1–17:
 electric power generation, 1–7;
 wind energy, 4, 6;
 solar energy, 4;
 hydroelectric/hydropower, 5–7;
 geothermal energy,
 5; biomass fuel, 5;
 U.S., 1–10;
 renewable futures, 8–10;
 non-U.S. overview, 11–17

Renewable futures, 8–10

Renewable portfolio standards (RPS), 8–10

Renewables obligation (United Kingdom), 12
Research and development, 22, 29, 34–35, 65, 71, 137–139:
  solar power/PV, 22, 34–35;
  teaming, 29;
  wind energy, 65, 71;
  geothermal energy, 137–139
Residential market (fuel cells), 193–194
Retail ready systems (solar/PV), 30–32
Run-of-river plant (hydropower), 158, 160

S

Sacramento Municipal Utility District (SMUD), 27
Sandia National Laboratories, 35
Santa Rita Jail (solar/PV case study), 41–42
Savannah River Site (bioenergy case study), 117–118
Shea Homes, 27
Shell Solar, 45–46
Shell Wind Energy, 70
Siemens Westinghouse, 185
Silicon solar battery, 40
SilvaGas process, 104–105
Siting (wind power facility), 71–74:
  hurdles, 71–72
Solar absorption cooler, 26
Solar basics, 21–32:
  concentrating solar power, 22–23;
  passive solar heating and daylighting, 23–24;
  solar process heat, 25–26;
  market for PV, 26–29;
  retail ready systems, 30–32
Solar Buildings Program (DOE), 34–35:
  research and development, 35
Solar cells, 21–22, 40–41: types, 40–41

Solar Energy Industries Association (SEIA), 35
Solar factoids, 39
Solar heating (passive), 23–24
Solar Home Solutions program, 31
Solar panels (solar/PV case study), 45–46
Solar Patriot house green design (solar/PV case study), 36–38
Solar power and photovoltaics, 4, 7, 14, 19–60:
  photovoltaics, 20–21;
  solar basics, 21–32;
  research and development, 22, 34–35;
  AESE engineering firm (case study), 32–34;
  DOE Solar Buildings Program, 34–35;
  Solar Patriot house green design (case study), 36–38;
  solar radiation types, 39;
  solar factoids, 39;
  solar cell types, 40–41;
  Santa Rita Jail (case study), 41–42;
  United Kingdom (case study), 42–44;
  Peru telecommunication systems (case study), 44–45;
  California State University Northridge solar panels (case study), 45–46;
  Peru solar grid-connect system (case study), 47;
  Pacific Park, Calif. (case study), 48–49;
  Learning From Light program (case study), 49–50;
  Parker Ranch, Hawaii (case study), 50–53;
  Glendale, Ariz. High concentration PV (case study), 54–57;
  Australia solar power tower plant (case study), 58–60
Solar power concentration, 22–23:
  parabolic-trough system, 22–23;
  dish/engine system, 23;
  power tower, 23
Solar power tower (solar/air flow case study), 58–60:
  financing, 59;
  design and construction, 60

Solar process heat, 25–26

Solar radiation, 39

Solar/wind hybrid system (case study), 50–53

Solid oxide fuel cell (SOFC), 176, 179–180, 182, 194

Son La hydropower project, 14

Sound impacts (wind power), 72, 77–78

South Africa, 16–17

South America, 15, 150, 155:
hydroelectric/water power, 155

Spain, 13, 66

Standard Market Design, 89

Steam, 131

Stelle, Illinois, 27

Storage facilities (hydroelectric/water power), 160

Storage plant (hydropower), 158, 160

Sulfur dioxide, 109

Sunspace, 24

Sweden, 85

T

Teaming (research and development), 29

Telecommunication systems (solar/PV case study), 44–45

Terminology, 205–232

Thin film solar cell, 41

Three Gorges Dam project, 14, 150

Transmission grid (U.S.), 88–91:
wind energy, 88–91;
rates, 90;
space availability, 90;
process, 91

Transportation applications (bioenergy), 114

Transportation fuels, 97–98, 114–117

Trombe wall, 24

Trust Fund for Renewable Energy and Energy Conservation, 142

Turbine technologies (hydroelectric/water power), 161–162:
Pelton turbine, 161;
Francis turbine, 161;
propeller turbine, 162;
Kaplan turbine, 162

Turkey, 16–17

TXU Electric and Gas, 80–82

TXU Europe, Ipswich, England, 42–43

U

U.S. Department of Energy (research and development), 22, 34–35, 65, 71, 137–139:
solar power/PV, 22, 34–35;
Solar Buildings Program, 34–35;
wind energy, 65, 71;
geothermal energy, 137–139

Ultraviolet (UV) light, 37

United Kingdom, 12–13, 42–44, 85:
solar/PV case study, 42–44

United States, 1–10, 20, 29, 63–64, 66–67, 88–91, 120, 125, 133–139, 148–149, 151–152, 188–190, 197–202:
renewable futures, 8–10;
wind energy, 63–64;
transmission grid, 88–91;
geothermal energy, 133–139;
hydroelectric/water power resources, 151–152;
industry contacts, 197–202

University of Nevada, Reno (geothermal energy case study), 139–142

UTC Fuel Cells, 186–187

V

Vertical-axis wind turbine, 62

Vietnam, 14

Visionary planning (wind energy), 78–79

Visual impacts (wind power), 72

W–Y

Wabash River energy facility, Indiana (fuel cells case study), 191–192

Water quality standards, 163

Western Europe, 12–13

Whole building design, 38

Wild and Scenic Rivers Act of 1968 (WSR Act), 168–169

Wind energy/wind power, 4, 6–8, 11–17, 50–53, 61–96:
wind/solar hybrid case study, 50–53;
wind farm, 62, 76–77, 82–86;
wind turbine types, 62–63;
wind resources in U.S., 63–64;
costs, 64–65;
research and development, 65, 71;
industry statistics, 66–71;
merchant potential, 71;
new technologies, 71;
siting hurdles, 71–72;
considerations of impacts, 72;
guidelines, 73–74;
land use, 74–75;
bird problem, 76–77;
sound concerns, 77–78;
visionary planning, 78–79;
Big Spring, Tex. (case study), 80–82;
offshore aesthetics, 82–86;
capacity and output, 86–88;
transmission issues, 88–91;
Italy (case study), 91–93;
Arklow Bank project, Ireland (case study), 93–95;
Brazil (case study), 95–96

Wind farm, 62, 76–77, 82–86

Wind Powering America, 65

Wind turbines, 62–63, 86–88:
types, 62–63;
capacity and output, 86–88;
speed, 88

Wind/solar hybrid system (case study), 50–53

Working fluid, 142

World overview (renewable energy), 1–17:
U.S., 1–10;
Mexico, 11;
Western Europe, 12–13;
Asia, 13–14;
Central America, 15;
South America, 15;
Eastern Europe, 16;
Former Soviet Union, 16;
Africa, 16–17;
Middle East, 16–17

Z

Zero-energy home, 36–38

Zimbabwe, 16